Countdown to Space War

Countdown to Space War

Bhupendra Jasani

Christopher Lee

sipri

Stockholm International Peace Research Institute

Taylor & Francis
London and Philadelphia
1984

UK	Taylor & Francis Ltd, 4 John St, London WC1N 2ET
USA	Taylor & Francis Inc., 242 Cherry St, Philadelphia, PA 19106–1906

British Library Cataloguing in Publication Data

Jasani, Bhupendra
   Countdown to space war.
   1. Space warfare
   I. Title.   II. Lee, Christopher, 1941-
   358′.8   UG1530

   ISBN 0-85066-261-3

Typesetting by Mathematical Composition Setters, Ivy Street, Salisbury, UK
Printed in Great Britain by Taylor & Francis (Printers) Ltd,
Basingstoke, Hants.

Contents

sipri

Stockholm International Peace Research Institute

SIPRI is an independent institute for research into problems of peace and conflict, especially those of arms control and disarmament. It was established in 1966 to commemorate Sweden's 150 years of unbroken peace.
The institute is financed by the Swedish Parliament. The staff, the Governing Board and the Scientific Council are international.

Governing Board

sipri

Stockholm International Peace Research Institute
Bergshamra, S-171 73 Solna, Sweden
Cable: Peaceresearch, Stockholm
Telephone: 08-55 97 00

Preface

President Reagan made a speech on 23 March 1983 that became known as his 'Star Wars' speech. In it, he called on scientists and engineers to find "the means of rendering nuclear weapons impotent and obsolete." The President was musing on the changes of using the considerable research into beam weapons and space systems as a counter to intercontinental ballistic missiles. Some dismissed Mr Reagan's speech as fantasy. However, what is not generally appreciated is the extent and the speed with which outer space has been put to military use.

Earth orbiting satellites are invaluable in the enhanced performance of armed forces and weapons. From 1958 to 1983 2114 military orientated satellites were launched. Seventy five per cent of all satellites have some direct military use. Their missions range from navigation, communications, meteorology and geodosy to surveillance.

Once the superpowers recognized the military value of satellites they immediately went about developing *anti*-satellite weapons.

At the same time, military satellites do have peaceful uses. They monitor potential enemies and therefore help maintain some form of uncertain stability. Indeed the major arms control treaties have clauses in them which allow for satellites to monitor areas where there could be violations of treaties.

Literature on such issues as anti-satellite weapons and space systems is starting to trickle from strategic and technical bodies. Yet most are written and presented for a reasonably informed audience. There is a need to reach a wider audience that does not have the technical nor the strategic understanding of the issues involved. That same technical and strategic understanding can be translated into simpler language and the issues explained. The authors, therefore hope that this book will help more people to understand the complex problems of the extension of the arms race to outer space and its control.

Acknowledgement

The editorial assistance of Gillian Stanbridge is gratefully acknowledged.

Bhupendendra Jasani
Christopher Lee

Stockholm
September 1984

1. Countdown to crisis

If the two superpowers go to war any time after, say, 1990, it is very likely that the war would start in space. The military use of outer space is so far advanced that by around 1990 both the United States and the Soviet Union will have sure means of damaging and perhaps destroying some of the satellites that would be crucial to commanders on either side. The Soviet Union may already have this capability. The United States appears confident that its current anti-satellite (ASAT) test programme will give it an ASAT weapon within the time scale that we are viewing.

The development of ASAT weapons started in the late 1950s. The need for them was apparent to the military in Moscow and Washington once it was seen that space flight was a reality and that the exchequers in both capitals were willing to support this modern form of exploration. Today, less than 30 years after the first Sputnik, the military commander has come to accept space as a legitimate environment into which he might extend his defence systems. About 75 per cent of satellites have some military application. They are used for spying, for sending and intercepting signals, as early warning systems and, in the case of the Soviet Union, are capable of hunting down and destroying spacecraft that could give an enemy too much of an advantage during wartime. It is not long since talk of orbital Big Brothers and hunter-killer satellites and beam weapons was considered to be the realm of fantasy. It was something for fireside reading and big-budget movies, stretching the imagination to the limit.

But, in 1983, Ronald Reagan spoke of a vision that modern technology could be developed to produce systems that one day might make ballistic nuclear weapons obsolete. He wanted American scientists to use their massive technological resources to build weapons that could knock out those missiles while they were still in space. There was talk of beam weapons, of lasers, of death rays. What Mr Reagan said on March 23, 1983, was soon built up into something that it was not. It became known as the 'Star Wars' speech—in short, it caught the imagination. The reality of it was that Mr Reagan's speech (in fact only part of a general address) was in 1983 nothing more than reflections. The systems to do what he was talking about did not exist. Nor do they today.

The problems of producing a ground-based or a space-based system that could pick up a massive invasion of ballistic missiles, lock onto each one and destroy them before they reached their target areas are enormous. The cynics have said that the problems are almost insurmountable. However, the research and development is under way. It is getting huge funding. The technology is emerging technology. It is growing in form and feasibility before the eyes of the space and chemical scientists and engineers. Furthermore, because of the way it has been sold to the public—as a protection against nuclear weapons and therefore against the concept itself— the Reagan vision of a futuristic ballistic missile defence system is gaining credibility.

This is especially so in the minds of those who have watched the space industry grow so rapidly since the first spacecraft in 1957. On top of all this there is the almost casual acceptance of these advances. The 'general public' finds no difficulty in accepting such a concept. Talk of beam weapons, lasers and spy satellites is so commonplace that there is an atmosphere that suggests that whatever people think of the ideas or whatever arguments there are against the principles, it will happen anyway.

Interestingly, many people who willingly accept the 'Star Wars' concept of being able to knock out ballistic missiles reject the principle that the superpowers should be allowed to knock out each other's satellites. Yet the two systems are linked, as we shall see later.

Unfortunately even among those who concern themselves with arms control and strategic studies, there is a lack of technical knowledge and understanding of just how and why space is being used today and what will inevitably follow during the rest of this decade. Even the peace groups have not added space studies to the growing portfolio of campaigns. There are problems of time and understanding. Society is only just coming to terms with the concept that wanting to know more about major military programmes is not a sign of pacifism nor of any one political persuasion. Furthermore, for those caught up in the seemingly endless debate on the nuclear systems here on Earth, space must indeed seem light years away in the importance of things. It is not.

The United States, for example, now has a branch of its Service called Space Command. It has its own generals, budgets and secretaries. Space is used for every essential function. Satellites have sensors that identify enemy positions, listen into enemy (and sometimes allied) communications and fix positions for ballistic-missile firing submarines. Others carry up to 80 per cent of all military signals.

Information is crucial to the military. Information about the enemy has to be collected. It has to be transmitted to the commanders who have to send information to command and control their own forces. The information systems used by the military are often brought together under the code name C^3I. It stands for Command, Control, Communications and Intelligence and relies heavily on space-based systems. A general or an admiral uses satellites to transmit *command* signals to his forces. He *controls* their movements by satellite. He receives their *communications* through satellites and it is the satellites that provide much of his general and sometimes detailed *intelligence*.

A major conflict, over a short period of time, relying to some extent on surprise and, initially, on rapid territorial and strategic advances, would need a great deal of management at every level. If the conflict lengthens, the problem of C^3I becomes even greater. It follows, therefore, that the commander's traditional intelligence needs remain unchanged whatever the type of conflict. He must have some good idea of his enemy's strengths, weaknesses, capabilities and readiness. He needs to know where his tanks are, what supplies have been coming through, whether aircraft have left their normal bases and dispersed to alternative airfields, whether there is an unusual amount of signal traffic, what reserves are mobilized and where they are being sent. His naval staff will want to know if ships are putting to sea in unusually large numbers, what types of ships are in the different task forces and what support vessels are going with them or being pre-positioned.

He will be interested to know if major towns are being evacuated, a sign perhaps that the authorities are preparing for air raids. He will wish to have a good idea of the lengths of his enemy's supply lines and how much activity is going on in them. Are train services, for example, being cancelled to clear lines for military railway units? The commander will even wish to know what the weather is like; the immediate weather reports and the medium- and long-term reports will be essential, especially to his air commanders, many of whom will not have aircraft capable of operating in bad weather conditions.

And if war does start, the commander will need constant updating on all these vital matters and will want it quickly. The way in which an advancing enemy division moves provides a good example of how an area picture might change during, say, a 24-hour period or even before an engagement. Most military formations are cumbersome and so are difficult to redeploy unless there is some extraordinarily well rehearsed procedure and excellent communications. The commander will want as much detail as possible about any division. How far ahead is the unit's reconnaissance group? How far behind are such units as chemical warfare detachments? Are the artillery groups also well forward so that they can be quickly deployed? How is the enemy protecting the advance, by putting out patrols on the route's flanks?

He will also need information about his own units. This may sound far-fetched; after all, a commander should be precisely briefed on where his own men and equipment are. But, in the preparations for a large conflict this may be quite difficult. For example, reinforcements may be experiencing all manner of problems joining up and supply lines may be broken through transport difficulties even before a conflict begins. The commander will need to talk to his own side, but securely, and he will need to get around the probable jamming put out by an enemy.

From all this it is clear that the military is getting to the point where commanders would be struck deaf, dumb and blind should their satellites be destroyed. Hence the added concern among many scientists and strategists: they believe that some future conflict could start in space. More importantly, many believe that, because of the military reliance on space systems, a war could start which otherwise would have been avoided by diplomatic means. The argument runs that in time of tension one side might decide that, to retain any military advantage, anti-satellite weapons should be used. That would be seen as a signal that diplomacy was about to fail and so war would follow.

Having said all this, it should not be thought that the only reason for space exploration is to satisfy the military. The peaceful programmes are important. Furthermore, both military and civilian satellites with their ability to look down, record and transmit in great detail what they see on Earth have great advantages for mankind. Apart from the obvious geological, cartographical and weather-forecasting advantages that satellites bring to every-day life, they can be used to monitor peace agreements. The same satellite that tells the American intelligence community that the Soviet Union has deployed a new missile could be used to tell if an arms control treaty is being broken.

The space age has been with us for some time. It is out of its infancy. In ancient times bands of warriors learned the advantage of sending a scout forward and high to spy out the enemy territory. They learned the advantage of taking the high ground. Space is the new high ground.

The first satellite was launched in October 1957. That was the beginning of the countdown to the crisis in space the world now faces. If space has a crossroads, it has now been reached.

We have a choice: to take control of this high ground in such a way that it will safeguard the peace of the world, or at least make some contribution to that peace; or to refuse to look up to this new and awesome responsibility.

2. How it all began

It all started in 1957. Very simply, the Soviet Union went ahead and did exactly what the United States had been talking of doing. On October 4, the Soviet Union launched Sputnik 1. The Sputnik (meaning satellite) took the US space programmers by complete surprise. In 1955 both the USA and the USSR had announced their intentions to launch small scientific satellites by 1957 as a contribution to the International Geophysical Year, but the Americans had no idea that the Soviet Union was anywhere near ready to launch ahead of them. And from Washington there were noises that indicated that the Soviet lead was more daunting for its political and prestige values than it was for its obvious scientific achievement. The message from as high as the White House was clear: "Let's get this particular show—not on the road—but off the ground."

But for those who envied the Soviet Union's propaganda victory, worse was to come. Within days of Sputnik 1's launch, work had been stepped up in the United States to bring forward its own plans for a space launch. And for three weeks after that launch Sputnik 1 sent out a steady stream of signals as a reminder of the Soviet Union's lead. Then there was silence. Sputnik 1 was still up there but not transmitting. Work continued on the US Navy's space project at the American Eastern Test Range at Cape Canaveral, Florida. But on November 3 the Soviets did it again. Sputnik 2 was launched from Tyuratam in Kazakhstan and that almost mocking bleep was heard for the next week over the United States. To make it even more interesting, Sputnik 2 had a passenger, Laika the dog (Laika means husky in Russian).

A month later, the Americans were ready. The Navy's Vanguard launcher was on site and the countdown going ahead as planned. On December 6, 1957, just two months after Sputnik 1, the Americans were ready for lift off into the marathon space race between the two superpowers. But it was a false start. The Vanguard launch vehicle lost its thrust after just two seconds and the spacecraft failed to go

6

into orbit. It was nearly two months before the United States scientists were successful. On January 31, 1958 Explorer 1 was sent into orbit and its signals monitored until they died out five months later, but not before it identified a belt of natural radiation about the Earth known as the Van Allen belt. The satellite itself continued in orbit until early 1970 when it decayed.

Although the Soviet Union had the early start and lead, its programme of space launches appeared to be limited. Certainly during the next few months and years the Americans were the more active. But in spite of the urgency fostered by prestige and concern that the Soviet Union might be gaining some military advantage, the programme could not be accelerated much faster than the existing schedule. Even then, the Americans were not always successful. For example, between Sputnik 1 in October 1957 and the first landing of a Moon probe in September 1959 there were 35 space launches. Nineteen of them failed—all 19 were American. The Soviets did not have such an active programme: only five of those 35 launches were from the Soviet Union. Sputnik 1 was the first in orbit; Sputnik 2 had Laika on board; in May 1958 Sputnik 3 went into space for two years, sending back signals until it decayed in April 1960; and in January 1959 the Soviets launched their Moon probe, Luna 1. Luna 1 passed within 6 000 km of the Moon and is now in solar orbit. In September of that year, less than two years after the first space shot, the Soviet Union launched Luna 2, again from Tyuratam. On Earth, space engineers held their scientific breath and then, 34 hours after lift-off, Luna 2 landed on the Moon.

Undoubtedly the Soviet success rate made it more difficult for the USA to live with its extensive failures. Perhaps an added problem for the Americans was their habit of telling the world exactly what they were planning to do—and when. Consequently the world was there to watch their failures as well as their successes, while it was and is felt that the Soviet Union's instinctive sense of secrecy and trait of publicizing only successes allowed them to gain some political as well as scientific credit. Whatever the rights and wrongs of handling the more public side of space exploration, the USA sometimes suffered excruciating growing pains during those first few years—especially in the rudimentary function of getting the satellite into orbit. Those early American failures were due to breakdowns in the various stages of the rocket engineering. The conclusions of the inquiries make disheartening reading: "Failed to orbit—control system malfunction", "Failed to orbit— unsuccessful fourth stage ignition", "Failed to orbit—third stage malfunction", "...third

7

stage ignition unsuccessful". "... second stage propulsion malfunction", and so it went on.

In February 1959, the USA launched its first military satellite—a photographic reconnaissance satellite. But it did not appear to attain orbit. Seven more Discoverer photoreconnaissance satellites were launched in 1959. Two of these failed to orbit and not one was successful in its mission. Even when the satellites did get into orbit, there were problems with the recovery of the film capsules. For example, about one month before the Soviet probe hit the Moon, the Americans used a Thor-Agena rocket to launch Discoverer 6 from the Western Test Range at Vandenberg Air Force Base, California. All went well at first; then the capsule was ejected on the 17th orbit but the recovery failed.

Table 1. Some major space firsts[a]

| Achievement | Soviet Union | | United States | |
	Date of launch	Designation	Date of launch	Designation
Artificial Earth satellite	Oct 1957	Sputnik 1	Jan-Feb 1958	Explorer 1
Animal in space	Nov 1957	Sputnik 2 (dog Laika)	Nov 1961	Mercury-Atlas 5 (chimp Enos)
First known photographic reconnaissance satellite in orbit	Apr 1962	Cosmos 4	Apr 1959	Discoverer 2
Unmanned spacecraft lands on the Moon	Sep 1959 (date of landing)	Luna 2	Apr 1967 (date of landing)	Surveyor 3
Anti-satellite system; Cosmos 185 possibly the first Soviet ASAT	Oct 1967	Cosmos 185	Oct 1959	B-47 aircraft-launched anti-satellite missile
Meteorological satellite	Apr 1963	Cosmos 14	Apr 1960	Tiros 1
Navigation satellite	Dec 1970	Cosmos 385	Apr 1960	Transit 1B
Early warning satellite	Dec 1968	Cosmos 260	May 1960	MIDAS 2
Communications satellite	Aug 1964	Cosmos 41	Oct 1960	Courier 1B
Man in space	Apr 1961	Vostok 1 Gagarin	May 1961	Mercury-Redstone 3 Shepard
Nuclear power source in space	Dec 1967	Cosmos 198[b]	Jun 1961	Transit 4A (Plutonium-238 fuel)
Electronic reconnaisance satellite	Mar 1967	Cosmos 148	Feb 1962	US Air Force satellite
Geodetic satellite	Feb 1968	Cosmos 203	Oct 1962	ANNA 1A
Woman in space	Jun 1963	Vostok 6 Tereshkova	Jun 1983	Challenger F-2 Ride
Nuclear explosion detection satellite; it is difficult to identify which of the Soviet satellites perform this mission	–	–	Oct 1963	Vela 1

Table continued

8

Table 1. Continued

	Soviet Union		United States	
Achievement	Date of launch	Designation	Date of launch	Designation
Accident involving nuclear power sources	Jan 1978	Cosmos 954 (carried a 40 kW(e) reactor)	*Apr 1964*	*A navigation satellite (carried plutonium-238 fuel)*
Fractional orbital bombardment system; USA does not have such satellites	*Sep 1966*	*Cosmos V.1*	–	–
Ocean surveillance satellites	*Dec 1967*	*Cosmos 198[b]*	Apr 1976	NOSS 1
Man on the Moon	–	–	*Jul 1969*	*Apollo II Armstrong and Aldrin*
Successful landing on distant planet	*Aug 1970*	*Venera 7 (landed on Venus, Dec 1970)*	Aug 1975	Viking 1 (landed on Mars, Jul 1976)
Manned laboratory	*Apr 1971*	*Salyut 1*	May 1973	Skylab
Re-usable spacecraft	Jun 1982	Cosmos 1374 (sub-scale model unmanned)	*Apr 1981*	*STS-1 Young and Crippen*

[a]Italics are used to indiacte which of the superpowers' achievements was the first of its kind.
[b]Cosmos 198 was the first of a series which may have carried nuclear reactors and which were moved into higher orbits probably to allow most of the radioactivity to decay. The USA has so far launched only one reactor into space.

These early problems were not confined to the American programme, although the United States did have a high failure rate partly because of the intensity of their launch schedule. Moscow's next venture into space, after the Luna 2 Moon shot, was the prototype of the Vostok series. This was the fourth Sputnik and at first the significance of the launch and what happened was not entirely appreciated.

Sputnik 4 went into orbit on May 15, 1960. Again the launch site was Tyuratam; what was very different was the mass of Sputnik 4. It was about 10 times the mass of anything else (except Sputnik 3) that had been sent into space. The first Sputnik satellite had weighed 84 kg. Satellites launched by the US Vanguard launcher weighed only 2 kg—considerably less than Laika the dog. Four days after the launch the Soviet mission control attempted to recover the capsule, but instead of coming down, it went up and into a higher orbit. The recovery failed.

It is important to realize that the size of Sputnik 4 indicated that these Soviet spacecraft were intended to be more than just transmitting instruments. Sputnik 4 was the prototype for a manned spaceship. Indeed, three months later the Soviet Union launched another large satellite, Sputnik 5. It stayed up just one day and on the 18th orbit the capsule was recovered. Inside were two dogs, Belka (Squirrel) and Strelka (Pointer). On December 1, 1960, up went Sputnik 6. The Soviets again tried a 24-hour recovery but failed and what was believed to have been a canine passenger was lost.

Meanwhile the Americans were pressing on with their programme in spite of some extensive failures, but with some successes also. 1960 was an important year in their military use of space (see table 1). The USA launched the first navigation satellite in April, the first early warning satellite in May, and the first communications satellite in October. In August, the Americans made their first ocean recovery and first mid-air recovery of a film capsule from a reconnaissance satellite. This was to prove important to the military at a later stage as a technique of getting back urgent high-resolution reconnaissance pictures ejected by satellites. That year NASA (the National Aeronautics and Space Agency) launched two weather satellites, a type later to prove useful to the military.

The same year, the Soviet Union apparently made two attempts to send probes to Mars. No official announcement was made as to the outcome of either of those flights, both launched in October. But by 1962 the Americans were suggesting that both shots had failed. This was of small comfort to the Americans, whose Moon probes launched on September 25 and December 15, 1960 failed. The first landed in Africa and the second exploded at an altitude of 13 km.

In February of the following year the Soviets had another go at long-distance space travel. Sputnik 7 is thought to have been a Venus probe, but decayed after about three weeks in flight. On February 12, 1961 two satellites, Venus 1 and Sputnik 8 were launched together on one launcher. The former, a Venus probe, did not go to the planet, but went into a solar orbit. Sputnik 8 decayed 13 days after being put into Earth orbit, which is quite common for Sputnik satellites. It is not clear how successful Venus 1 became as a prototype for long-distance exploration.

Still the Americans were having troubles. Between Febuary 1961 (the month of the Soviet Venus probe) and April 8 the USA launched a series of spacecraft which malfunctioned in some way, usually because of stage motors failing or capsules failing to eject. In the Soviet Union at Tyuratam, the Soviets resumed the heavyweight Sputnik series. Sputniks 9 and 10 both had dogs on board. It was only a matter of time before a cosmonaut—an air force pilot—was introduced. It happened on April 12, 1961. The big Sputnik series changed names and Vostok 1 was sent up and brought back on the same day after orbiting the Earth once. On board was Yuri Gagarin.

Three weeks later the Americans replied by starting their Mercury series of manned flights. On May 5, 1961, Alan B. Shephard went into space although not into orbit. His sub-orbital flight lasted just a little more than 15 minutes, as did that of Virgil I. Grissom the following July. The Mercury experiment continued with practice splashdowns, and the inevitable failure: one unmanned satellite failed to get into orbit and was destroyed by mission control's range safety officer.

On November 29, 1961, the Americans put a chimp called Enos into a spacecraft, sent him into two orbits and recovered him safely. On February 20 of the following year it was the turn of John Glenn in Mercury-Atlas 6; his was the first American orbital flight in space and to some extent it was a turning point or a catching-up point for the Americans. True they suffered major failures, but by 1962, just five years after Sputnik 1, it was clear that both superpowers were regarding space as a very important scientific and military playground in which they were both willing to invest huge financial and technological resources.

Others too wanted to invest in space. British, French and Chinese scientists and their governments were particularly interested and there was already the recognized need to establish a European space agency on the basis that no one European country could really afford to maintain a single comprehensive programme. By 1962 there were joint programmes in hand with the United States. The British joined with the Americans to launch a satellite to study the ionosphere in April of that year under joint US–UK directorship. By September the first country outside the superpowers to direct its own project was ready and on September 29 the Canadian satellite Alouette I went into orbit about 1 000 km above the Earth, launched by the USA.

That year marked the start of a heavy programme of launches, especially those from the American sites. Successful launches included two military firsts—a geodetic satellite and an electronic reconnaissance satellite (see table 1). During 1962, 68 spacecraft were sent aloft successfully; with some fluctuations it was to set the pattern for crowded schedules until the end of the decade (see figure

1). Of the 68 launches that year 52 were American or American launched, and of these 25 were US Air Force photographic reconnaissance satellites. As well as the manned craft of Glenn in February, Carpenter in May and Schirra in October, the Americans continued with their long-range projects, including probes to the Moon and beyond. In July they tried a Venus probe with Mariner 1 which failed and was remotely destroyed at an altitude of 160 km. Mariner 2 did get off safely in the late summer and flew past Venus on December 13, 1962. The concentrated American programme inevitably meant a continuing high failure rate; put another way, in the five years after the start of the American programme with the

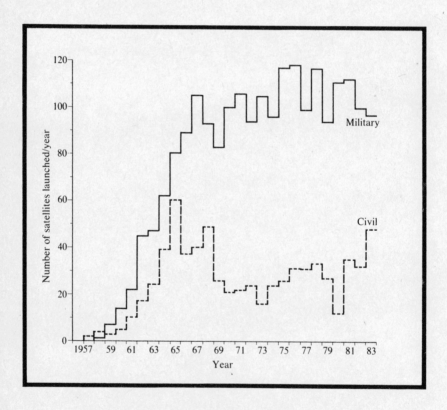

Figure 1. This chart of successful military and civilian satellite launches by all countries shows when the superpowers mastered the art of launching spacecraft. By about 1970 they had passed the experimental stage and entered the routine period as indicated by the roughly constant numbers of launches yearly.

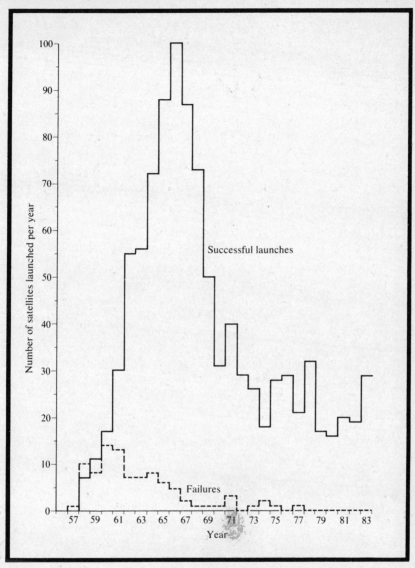

Figure 2. These days unsuccessful launches are a rarity. This figure shows a comparison of successful and unsuccessful US space launches. Owing to lack of knowledge of the Soviet failures, similar analysis of Soviet launches is not given.

US Navy's ill-fated Vanguard flight on December 6, 1957, 52 per cent of the American launches were successful (see figure 2).

The Soviets launched only two manned flights that year, Vostok 3 and Vostok 4. Between August and November 1962 they made six attempts with Venus and Mars probes—or at least it is thought that they made six. One of the problems of analysing Soviet space programmes was and is the scarcity of published material from Moscow. Of the six long-range shots, five appear to have failed. The one launched on November 1 was probably successful and the craft is now in solar orbit. But the importance of 1962 for the Soviets was that it saw the start of the Cosmos series.

Cosmos 1 was launched on March 16 from a new space centre at Kapustin Yar on the Volga River, using a new type of launch vehicle, a Sandal intermediate-range ballistic missile augmented by an upper stage. Until this point the Soviet Union had used what has been designated the A series of launch vehicles, which was based on the first Soviet intercontinental ballistic missile (ICBM), the SS-6 and given the NATO codename Sapwood. Two months after the Soviets tested their new SS-6 missile it was used to launch Sputnik 1. All the A series of rockets were fired from Tyuratam.

With the start of the Cosmos series came an announcement from Moscow of a comprehensive programme of scientific research for the Soviet space programme. However, it would not be unreasonable to point out that the Soviet programme had always had a largely military basis, as has been the case with the USA, both countries using mainly Air Force men as astronauts/cosmonauts.

1963 was not a remarkable year for space travel in the purely scientific sense. One of the important long-term projects known as Dyna-soar was cancelled in December of that year. The Americans had produced Dyna-soar as a design for a one-man winged spacecraft that could not only manoeuvre but also land. To many the concept was fascinating and seemed a logical step in creating a spacecraft that could be used again and again. But it remained an experiment and the programme was dropped, because, not for the first time, developments overtook experiments.

In its place came the MOL, Manned Orbiting Laboratory. The system was to have been a cylindrical laboratory attached to a Gemini spacecraft. The whole thing would be put into space by a large Titan rocket. It never got off the ground; once again, the programme was overtaken by developments. If it had, it would have housed a then-sophisticated monitoring and reconnaissance unit. One of the reasons that it was dropped from the programme was that many of these military tasks could, by 1969, be carried out by unmanned spacecraft. This was, and might well remain, a reminder that space technology advances should never be underestimated.

One of the most significant developments occurred in November 1963 when the Soviet Union launched Polyot I. This was the first spacecraft with the capability to manoeuvre extensively and this was seen as an important advance in man's ambition to 'use' space rather than be governed by its uncertainties. It also foreshadowed the development of the hunter-killer satellite, for which the ability to manoeuvre is, of course, a basic requirement.

The previous month the Americans had launched the first two of their satellites designed to monitor nuclear explosions. These were Vela 1 and Vela 2. The theory was simple: the satellites would go into near circular orbits at an altitude of more than 100 000 km (see figure 3, orbit E). They would not be together but on opposite sides of the Earth, the idea being that they would in effect give 24-hour cover of the Earth, supposedly able to detect any atmospheric explosions as well as those carried out in outer space. Each satellite would be equipped to pick up signs of gamma radiation, neutrons and the characteristic light flash of an atmospheric nuclear explosion: these signs would be transmitted to a monitoring base on Earth. Because of their high orbits, it has been estimated that the Vela satellites have an orbital life of some one million years, although of the 12 launched only 2 are now working.

At lower altitudes between 150 and 500 km, the US Air Force put up 21 military photoreconnaissance satellites (although 4 failed to get into orbit). The Soviets sent up seven similar craft orbiting between 190 and 400 km using mainly low-resolution cameras.

The resolving power of a sensor is the minimum distance between two small objects when they can still be distinguished as two separate objects. Normally, with a high-resolution space sensor objects between 150 cm and 2 m in diameter can be distinguished, whereas a low-resolution sensor can distinguish objects between 3 m and 5 m in diameter.

However, the publicists of space exploration now had what they had been waiting for ever since Sputnik 1: the first woman in space. Valentina Tereshkova was strapped into Vostok 6 on June 16, 1963 at the Tyuratam launch pad. Two days earlier, Vladimir Bykovsky had been sent into orbit in Vostok 5. Bykovsky was still in orbit when Tereshkova was launched and the two passed within 5 km of each other more than 100 km above the Earth. They both landed by parachute on June 19, Bykovsky after 81 orbits and Tereshkova

15

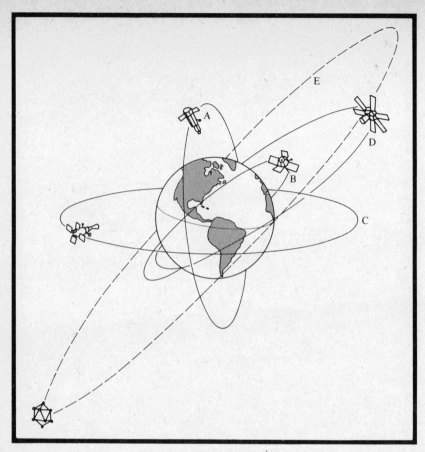

Figure 3. Various types of satellite orbit; characteristics of the various orbits are given in the table opposite.

after 48. There is a story told in the Soviet Union of how Valentina Tereshkova had but a rudimentary knowledge of space flight and furthermore suffered certain discomfort during her 70-hour journey. It is said that when she landed it was beyond the scheduled dropping zone and that she was found wandering in a distressed state by an old woman in a remote part of the country. It is said also that the woman was concerned and confused by Tereshkova's appearance and insistence that she had come from "up there" and her frequent references to a seagull. The woman is alleged to have comforted her with soup and blankets. She was sure she had come across a mad woman in need of humouring and the firm hand of the authorities, not immediately realizing that she had found one of the lasting

16

Figure 3 (continued)

Satellite	Altitude (km)	Orbital inclination[a]	Type of orbit
Reconnaissance			
Photographic			
Big Bird	180–290	97 °	A
KH-11	240–530	97 °	A
Cosmos	180–350	62 °, 72 °, 82 °, 67 °	B
Electronic			
US	480	97 °	A
Cosmos	500	74 °	B
	650	82 °	B
Ocean surveillance			
US	1 100	63 °	B
Cosmos	250	65 °	B
Early warning			
US	36 000	0 °	C
Cosmos	688 × 39 000	63 °	D
Nuclear explosion detection			
US	110 000	35 °	E
Meteorological			
US	36 000	0 °	C
Meteor	900	81 °	B
	610	98 °	A
Communications			
US SDS DSCS	250 × 39 000	64 °	D
US FLTSATCOM	36 000	0 °	C
Molniya	440 × 40 000	63 °	D
Cosmos	1 400	74 °	B
Navigation			
US	20 000	64 °	B
Cosmos	1 000	83 °	B

[a]The angle between the plane of the orbit and the equatorial plane.

heroines of the Soviet Union and somebody who would be held high as an achievement of Soviet womanhood.

During 1964 the Americans abandoned their manned flights, having completed their Mercury programme the previous May, but they did launch the third and fourth nuclear explosion monitors, Vela 3 and Vela 4. They also had their first success with the Ranger series of Moon-shots, getting back more than 4 000 pictures as these craft orbited the Moon or crash-landed; and they continued with the preliminary testing of what were to become the successful Gemini and Apollo series.

The Soviets experienced failure with their Venus and Mars probes but success with the first three-man crew in space: Voskhod 1 made 16 orbits in October. The following year, 1965, Voskhod 2 went up and this flight included a space walk. But it was to be the last Soviet manned flight for two years, when the Soyuz (Union) series started.

The Americans, however, kept their manned programme going. During March 1965 Gemini 3 with Grissom and Young aboard carried out the first manned manoeuvres in orbit. Gemini 4 and Gemini 5 flew later that year, and Geminis 6 and 7 came within 30 cm of each other in a complicated but successful rendezvous operation.

1966 opened with a flurry of January launchings, with one a failure. One success that month was the Soviet Union's Luna 9. It took three days to get to the Moon and then for another three days sent back pictures of the surface. But, two months later, what appeared to be a Moon probe, Cosmos 111, failed after two days in a low-Earth orbit. March was a good month for the Americans, who were gaining more confidence with each flight (see figure 2). Gemini 8 was used for the first space docking test although because of some electronic fault the complete flight programme was cut short and the astronauts Armstrong and Scott returned safely to Earth.

Later in the summer there was a less spectacular but equally significant event. The Americans launched what was then called IDCSP—Initial Defense Communication Satellite Program. Seven military satellites were launched using a single Titan rocket. This was the forerunner of the system that allows military commanders to communicate between different elements in their forces, ships, planes and ground units. The military remained convinced that this communications role for space technology was essential to its future war planning.

Later that year a further eight satellites were loaded into another Titan rocket. This time however the Titan broke down and the important package failed to get into orbit. The previous year the Soviet Union had launched its Molniya (meaning lightning) series of communication satellites—including a tie-up with the French. There have been suggestions that the Soviet military commands used this system in the early days of space-based communications.

Another important event during 1966 was an attempt by the Japanese to launch their own test satellite from a site at Kagoshima on the island of Kyushu. The flight was a failure, but it was a reminder that space travel could not be considered the prerogative of countries such as the USSR, the USA and the leading Europeans. Indeed, later the same year the Japanese tried again. It was another failure and there were more to come, but in February 1970 the

18

Japanese did succeed in launching their own engineering test satellite and so did the Chinese.

Meanwhile the Soviet Union started a programme that was not immediately understood by Western space watchers. From 1965, Western military attachés reported that the Soviet Union had developed what was described as an orbital missile, the SS-10. Nobody quite knew what to make of this information. But the military and scientific detective work started in earnest in the late summer of 1966. On September 17 the Soviet Union launched a satellite and for the first time did not announce its entry into orbit. And then, whatever was being carried—the payload—was exploded. At the Earth-based monitoring unit the Americans detected debris from the explosion but could not explain it. Then, in November, a similar satellite was launched. Once again there was no announcement from Moscow (four other launches between September 17 and November 2 had been announced in the usual manner). Once again the payload was exploded and debris detected. It has been assumed that the Soviet Union was carrying out its first test of a FOBS (Fractional Orbit Bombardment System).

The idea behind a fractional orbit bombardment system is simple. A weapon (possibly nuclear) is sent into orbit, but before that orbit is completed, at a given point a retro-rocket is fired. This slows down the weapon and it is then dropped onto its target on Earth. It is thought that the Soviet Union tested this system of bombing from space between 1966 and 1971 with the most concentrated period coming during the second half of 1967.

1967 was a sad year for space exploration.

The Soviets had not flown any manned craft for two years and the Soyuz programme was scheduled to start in 1967. On April 23 Soyuz was launched as planned. On board was the cosmonaut Vladimir Komarov, who in 1964 had flown with Yegorov and Feoktistov in the first Voskhod flight. After nearly 27 hours in orbit a recovery was attempted but the parachute became entangled and Komarov was killed. There were no further flights from the Soviet Union until October of the following year.

In the United States work was going on to test the basic systems in the planned Apollo series. On June 27 three astronauts, Gus Grissom (the second American to go into space in 1961), Ed White (in 1965, as one of Gemini 4's crew, he had space walked for 21 minutes) and Roger Chaffee, were killed in a fire on the launch pad. They had been scheduled to be one of the three-man crews preparing for the Apollo series; and it was not until October 1968 that the manned Apollo programme began with Schirra, Eisele and Cunningham. That flight lasted for 163 orbits and 260 hours and it was later described as "near-flawless".

In between the 1967 disasters and the success of 1968, the Americans concluded their Surveyor Moon-shot programme with only two failures. This programme had included the first landing and rocket take-off from the Moon (Surveyor 6). Meanwhile the Soviets were continuing their FOBS tests as well as a 128-day flight to Venus. That was June 1967 and it was the prelude to a double Venus probe during 1969 when the Soviet spacecraft Venera 5 and Venera 6 entered the Venusian atmosphere within a day of each other and presumably landed, although with what precision it is difficult to tell from published data. This was overshadowed by the big event of 1969.

On July 20, 1969, Eagle, the lunar module from Apollo 11, landed on the Moon with Neil Armstrong and Buzz Aldrin. The giant step for mankind had been made just 12 years after Sputnik 1. On November 14 of the same year Apollo 12 was launched from the Eastern Test Range at Cape Kennedy (Cape Canaveral), Florida, and five days later its lunar module Intrepid with Conrad and Bean on board landed in the Ocean of Storms (Eagle had landed in the Sea of Tranquility). Space travel was advancing at a furious rate. During that November, Britain's military communications satellite, Skynet, was launched by the USA and positioned over the Indian Ocean. Three months later, Japan achieved its independent launch at long last and in March 1970 the French launched a Franco-German satellite. This was the first foreign payload launched by the French and served as a further reminder, along with the Japanese, that space-travel capabilities were spreading.

Ten days after the Franco-German launch, NATO's joint military communications satellite was launched by the USA and on April 24 China sent up its first satellite. Towards the end of the year, the French put up a weather satellite and, of particular interest, the Italians joined the space business. They became the first to launch a satellite for the Americans. The United States had a small astronomic system called SAS 1. It was this craft that the Italians launched from San Marcus using a US launcher.

In general terms, however, 1970 will be remembered as the year the world held its breath for what could have been the most public disaster in the short history of the space age. Appropriately named Apollo 13 was launched on April 11. Fifty-six hours later, while still on the way to the Moon, there was an explosion and the Number

2 oxygen tank was wrecked. The decision was taken to scrap the mission to land on the Moon; but the problem remained how to get the men back. On board were astronauts Lovell (the most experienced of all American astronauts), Haise and Swigert.

The first problem was to assess the damage and to see what else might be damaged by it. Then it was necessary to check both on-board systems and those connected to mission control at Houston. The crew flew their damaged spacecraft round the Moon and headed towards Earth. For more than three and a half days the world watched as mission control and Lovell, Haise and Swigert went through every procedure possible to assure their safe return. And then, 86 hours and 53 minutes after the malfunction, the Apollo capsule splashed down in the Pacific. Strangely, what could have been a terrible event appeared to win the public's confidence that the technology was advanced enough to cope with major difficulties faced by men thousands of miles from the accident-prone six-lane highways of North America.

Apollo 13 did not slow the general lunar programme for the Americans, who were more advanced than the Soviet Union (although the Soviet Union landed Moon probes, there was never a manned flight to the Earth's natural satellite by the Soviets.) During the next two years, the Americans sent four three-man crews to the Moon in Apollos 14, 15, 16 and 17, with no further mishaps. The American manned flight programme then stopped until May 1973, when it resumed with the four 1973 flights of Skylab.

Skylab was more or less what its name suggested: a laboratory in the sky. It weighed 74 783 kg and housed a multiple docking facility, a workshop, a special telescope mount and of course the facility to keep men in space for some considerable periods, although the Americans never attempted the marathon space flights of the Soviet Salyut cosmonauts (see figure 4).

Salyut 1 had been launched in April 1971. It was nothing more than a prototype for a manned space station. Although the Soviet space effort continued to look further afield towards the planets, there was every sign that they would concentrate their efforts in near-Earth exploration. With the military in Moscow having such a say in the programme, this was only natural and at that stage of their programmes they were limited by their lack of micro-technology, which would have made long-distance travel that much simpler.

When Soyuz 11 docked with the Salyut 1 space station in June 1971 there were those who believed that the Soviet Union had once

21

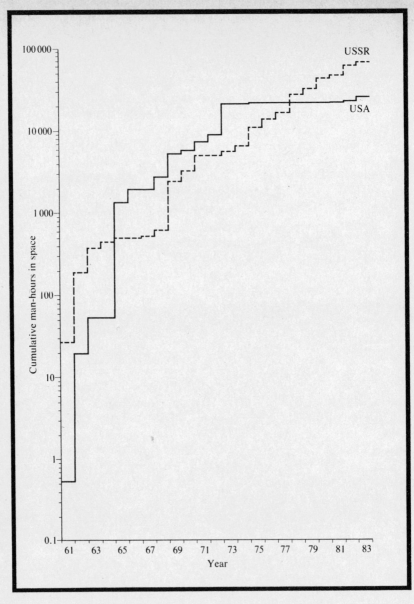

Figure 4. One aspect of the space race: the total number of hours spent by astronauts and cosmonauts in space. In this respect the USA overtook the USSR in 1965; only in 1978 did the USSR catch up and pass the USA.

again proved to be ahead of the Americans by getting into the space station business ahead of their rivals. But for the Soviets that space success was to end in great sadness. The three cosmonauts on board, Dobrovolsky, Volkov and Patsayev, went through their prepared programme without any problems and then moved back into Soyuz 11 for their return journey to Earth. When Soyuz 11 was opened, the three cosmonauts were found to be dead. They had apparently died of asphyxiation after the separation from the space station. There were no further Soviet manned flights until 1973.

1973 was a busy year, not in terms of numbers but in practical use of satellites. It started with the Soviet Union landing an un-manned craft on the Moon and sending out its Lunakhod rover. It included, in April, the American Pioneer 11 being launched as a Jupiter probe; the following month the first three-man crew went aboard Skylab orbiting more than 400 km above the Earth. In June the Soviet Union launched eight military satellites with one rocket (the first octuple launch had occurred in April 1970). The satellites Cosmos 564 to 571 were put up as part of the military communications system. In August they had two successful Mars probes, as had the Americans during the previous month. But it was the autumn of that year that was to demonstrate yet again the worth of satellites to the military.

On October 6, 1973, the Yom Kippur war started. There had been tension in the area for some weeks and during the last few days of September it appeared to the superpowers that war was inevitable, although nobody could anticipate the extent of the fighting. The Americans had immediate satellite cover thanks to their longer-living spacecraft, and perhaps some diplomatic advantages. The Soviet Union's photoreconnaissance satellites for example have had the disadvantage (in Western opinion) of having an orbital life of only a few days before they are recovered for their intelligence. The Soviet Union has not published details of any kind that would indicate officially their activity during that autumn, but looking at the heights and angles and other orbital characteristics of a series of Cosmos satellites it is possible to make certain assumptions that point to one of the ways in which the Soviet Union monitored that war and its aftermath (see figure 5).

Three days before the fighting started Cosmos 596 was launched, it was somewhat restricted because it could not be manoeuvred; however it did supply what was needed at the time: an overall picture of the deployment of the various forces in the area. At that stage it was more important to get an area rather then a detailed

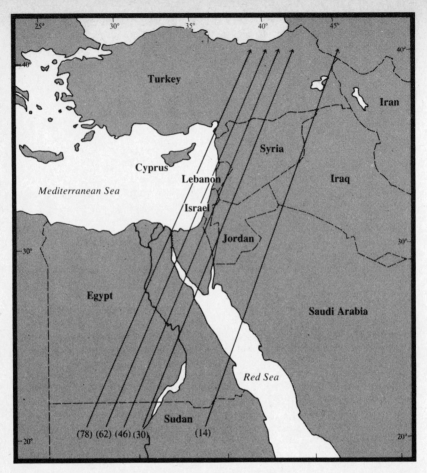

Figure 5. 'Eyes' in space. The superpowers' photoreconnaissance satellites often observe conflict areas on Earth. Here we can see the ground tracks—the satellite's path projected onto the surface of the Earth—of the manouevrable Cosmos 597 satellite over the Middle East in 1973. (The figures in brackets indicate the number of the orbits.)

picture of the military situation on the ground. Cosmos 596 stayed up for six days. It was recovered three days after the outbreak of the conflict for analysis. On the day that fighting began, Cosmos 597 went up (see figure 5). It had the advantage of being manoeuvrable and it was brought back on October 12. Cosmos 598 was launched on October 10 and recovered six days later; on October 15 Cosmos 599 went up, although its mission appears to have been two-fold

24

with much of its Middle East intelligence gathering carried out from about the 24th of the month. It was recovered on October 28. Cosmos 600 was launched on October 16 and brought back on the 23rd. The next Cosmos, 601, was not a military photoreconnaissance satellite; it probably monitored electronic signals. Cosmos 602 was a reconnaissance spacecraft as were seven others launched during the last two months of that year. During the next 12 months military satellites dominated the launching schedules.

The Soviets continued their programme of sending up military communications satellites in batches of eight on one rocket (Cosmos 617–624, Cosmos 641–648 and Cosmos 677–684). The Americans too maintained their military communications satellite programme; in 1974 they sent up their own reconnaissance satellites including a short-lived orbit during February and another in June. The American Big Bird series was already in flight when Turkish forces invaded Cyprus on July 20 (see figure 6). During that Eastern Mediterranean crisis, the intensity of the satellite reconnaissance rivalled that over the Middle East during the previous October (see figure 5).

While all this was going on, the Soviet Union kept up with its manned flights. These were in the Soyuz series and tied with the use of the Salyut near-Earth space station. The Americans had no manned missions after November 16, 1973 (Skylab) until the Apollo 18 flight of the veteran astronaut Thomas Stafford (who had first gone into space with the Gemini series 10 years earlier) and his crew of Slayton and Brand on July 15, 1975. The Americans then closed down their manned-flight programme until 1981 and the first space shuttle. There were those in the United States who claimed at the time that this was a major mistake on the part of the Administration, but others pointed out that there could be no real justification for such flights and that every effort should be put into getting the space shuttle in the air and into service.

The Soviet Union has a continuing manned programme that started in 1974 and has run without a break ever since (see figure 4). It included the establishment of the two important space stations, Salyut 6 and Salyut 7, when a series of space records for manned flight were made. The Salyut space station showed that man could live in space for extended periods without any long-lasting side effects, although there was some doubt about the operating efficiency of cosmonauts after as much as six months in space. One said on his return that it was possible to stay there for as long as the mission required. Some of the Soviet cosmonauts did experience unusual

25

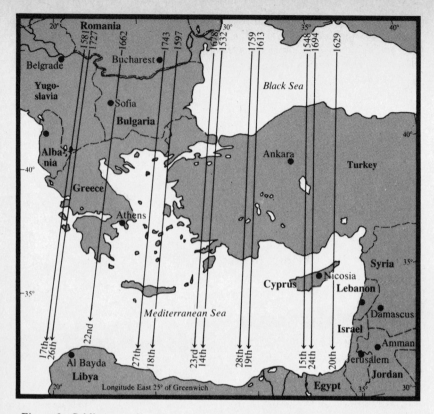

Figure 6. Soldiers may be seen from space way below in Nicosia, as the satellites of the superpowers train their cameras on another world trouble spot. This depicts the ground tracks over Cyprus, Greece and Turkey of the US 1974-20A satellite, launched on April 10, 1974 at an orbital inclination—the angle of the plane of the satellite's orbit to the equatorial plane—of 97.52 ° on July 14–28. The date and orbit number are indicated for each ground track.

effects in space. For example, one of them 'grew' about 8 cm but returned to his normal height after a few days on Earth subject to normal gravity.

A further series of biological tests were carried out in 1982, when the second woman in space, Svetlana Savitskaya, joined the Salyut series. It is understandable that in the male-dominated world of space travel there should be such curiosity when a female goes into space. (At the time of writing there have been but two Soviet women

26

and a single American, Sally Ride.) Savitskaya's role was certainly more managed, professional and scientific than that of the first woman cosmonaut, Tereshkova. It is generally accepted that Tereshkova's flight was ill-prepared and that neither she nor her ground support colleagues were fully aware of some of the problems that space travel may cause for women. Savitskaya's role was quite clearly defined by the Soviet Union's space officials once the flight was under way and after its successful completion. Nobody doubted her qualifications as an aviator test pilot and training as a cosmonaut, but the Soviets made it clear that it was important for her to carry out "certain biological experiments" and for this she was given separate accommodation.

During the Soviet Union's preoccupation with the Salyut space station series there were those who speculated that America was falling behind in the space marathon. However, on April 12, 1981 the United States proved its lead in one of the most advanced areas of space travel by successfully launching and recovering Columbia, the space shuttle. The importance, of course, was that it could be re-used and it could not only launch satellites but also retrieve them. For the first time, man really had got close to the science fiction comic strips that allowed spacemen to take off and land with the ease of fast jet pilots. Perhaps that is still some way off. Yet considering how far man has come since October 4, 1957, space flight with the minimum of ground support and simplified launch procedures is no longer such a fanciful concept.

Certainly there are sufficient interest groups to put the pressure on government to produce and maintain the funds to make this so. The Soviet programme has never given many signs of lacking funds and official enthusiasm. In many ways it is seen as an important annex to the Soviet Union's military programme. In Washington there have been many times when the American programme has been threatened by the demand for funds by other groups; when the Apollo programme ended there were many lobbyists who claimed that the United States space programme was defunct and that NASA was nothing more than a care and maintenance organization. Certainly the American military has had a major problem. On the one hand, it has to lobby for more defence dollars for conventional projects. On the other, it wanted a great say in the direction of the nation's space programme; and the only way to maintain that say was with a "dollar input".

The military has always had a major interest and role to play. The first American attempt to get into space in 1957 was directed by the

US Navy. (Eight of the 14 US space shots during the first year were Navy-directed flights.) And the United States has now established its own military space command and its own shuttle facility is being built at Vandenberg Air Force Base in California. But when it is remembered that about three-quarters of all space flights have either a direct or indirect military function, it is little wonder that the military should wish to have at least that percentage say in the future of space travel. Therefore it is worth, at this stage, looking at some of the space systems and trying to identify which satellite does what and for whom.

3. Space—the military domain

For most of us there remains a fascination for space exploration. More than a quarter of a century after Sputnik 1, the engineers, scientists, cosmonauts and astronauts are still frontiers-men and -women. In spite of the advances and demonstrations of remotely controlled spacecraft there remains among the public a great holding of the breath; it is almost as if space travel is the last of the Greatest Shows on Earth. True, once man had stepped on the Moon, international imagination and interest was less easily captured by the scientific and human daring that followed. But it would have only taken some tragedy in space to recapture that interest. If, for example, the disaster on the American launch pad that claimed three lives had taken place when the craft was in orbit, the next launch and recovery would have been watched with greater interest. This somewhat macabre view is used only to demonstrate how possible it is for governments to advance more or less any programme that they wish knowing that international opinion will be indifferent to its implications either because it doesn't care what is going on or because that same international opinion is overwhelmed by the technicality of the achievement. (There is in this argument also a hint of public fickleness.)

At the same time, society has surely reached a point when it might ask why so much skill, attention and finance is being devoted to space exploration. In the early days the answers were obvious; they had a great deal in common with the answer to the question: Why climb mountains?—Because they are there. Space was and remains a challenge. The first flights reflected more than a sense of remarkable achievement. Science and technology would have been shirking their responsibilities if they had not turned their attentions and resources to space exploration. Equally, the various departments of defence would have been naive if they had resisted the temptation to encourage and take advantage of this exploration.

Beyond the laboratories and government departments, the

29

average person would be hard pressed to produce a list of advantages to him or her of the billions of dollars spent since the different programmes started (see figure 7). Of course, there have been advances: everything from micro-technology development to instant communication via satellites (including those for that great home psychiatrist—the television), clothing, medicine and materials. The real advantages to mankind may be still to come and to many there is always that faint glimmer of hope and fantasy that one day space travel and signals may link Earth to some other form of comprehensible life.

Meanwhile most of that exploration is not on the edges of this galaxy millions of kilometres away, but just a few hundred kilometres above the Earth. There are many reasons for this. One is the technical limitations of what is still a young science. Another is equally obvious: there are not many places to send spacecraft. A third is that one function is to find out more about Earth and therefore an important role for a spacecraft is to look down, not up. That is a fact recognized by the military: the taking and commanding of the high ground. It was the late John F. Kennedy who offered the thought that: "if the Soviets control space, they can control Earth, as in past centuries the nation that controlled the seas dominated the continents".

If it had not been for the military, it would have taken longer for science on its own to get into space to start any form of exploration. The military had one technical advantage to offer. Public attention usually focuses on the satellite, the lunar probe, the landing module or the interplanetary explorer. But something has to boost them into orbit in order that they can start their journeys. In fact, as noted in Chapter 2, the biggest problem for the American programme was not the failure of the satellite, but the inability of the launchers to start the whole process.

The defence departments of both superpowers were the only organizations which had an original need for the launching power, the stage rockets necessary to get the scientific and military packages into orbit. Thus, the first successful American launcher was a military rocket and at least two others were based on similar vehicles.

Not only did the military have the wherewithal for launching objects into space, they also had, as we have pointed out, a very active interest in the uses of satellites. This interest had started back in the 1940s. The US Navy began its satellite study in 1945 and the following year the US Army Air Force launched its own feasibility study. That year both services talked about establishing a joint

30

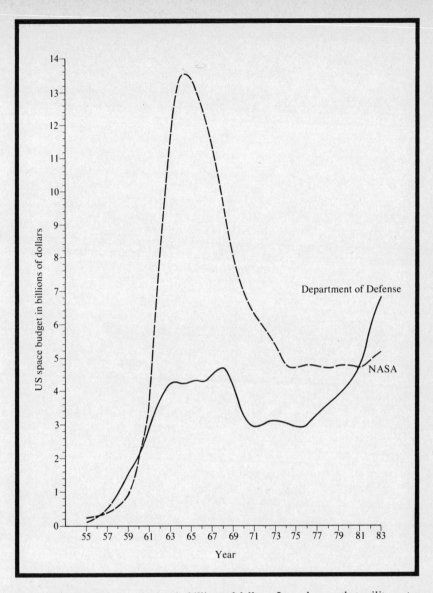

Figure 7. The US space budget in billions of dollars. It can be seen that military space expenditure has now overtaken civilian; during the 1960s manned space flight pushed up NASA spending. No equivalent Soviet figures are available. The expenditure has been adjusted to 1980 prices.

31

programme. But this initial idea came to nothing. (Joint service projects have had a habit of flopping as inter-service rivalries dominate what should be reasonable programmes. Too often basic disputes—such as which service is to chair the project, how funding might be split and where energies may be concentrated—stop the most obvious joint projects from getting off the ground.) Both the US Navy and the Army Air Force recognized the need to start a comprehensive space research programme but nothing really got going until the 1950s.

The Russians had toyed with rocketry since the 17th century. As early as 1829, rockets were used in the war between Russia and Turkey. In the same century, Konstantinov was publishing in Russia his thoughts about space flight. This was far more than fantasy for him and some of his rockets had a range of about four or five kilometres. Towards the end of the 1800s this concept was taken a stage further by K.E. Tsiolkovsky. He produced a detailed description of a rocket engine which he believed could be the basis for interplanetary flight. To some extent this was still early science fiction with a slide rule to back it up, but between 1903 and 1914 Tsiolkovsky was confirming his position as the father of Russian space engineering. It was during this 11-year period that he produced what many have accepted were sound principles of space rocket design. By 1929 he had recognized the need for a multi-stage rocket. But it was to be the next generation of rocket designers that was to take the Soviet Union closer to and beyond the edges of space. One group, for example, included S.P. Korolev, a particularly talented engineer; Korolev's work started in the 1930s and continued after World War II. It was his group that designed the Soviet Union's first intercontinental ballistic missile.

It was tested in August 1957 and, two months later it was used to send Sputnik 1 into space. For both the United States and the Soviet Union the importance of building rockets was essentially a military priority. Certainly, pure scientific curiosity recognized the importance and ambitions of space travel and it would be misleading to imply that space exploration has been a wholly military operation. But probably without the military requirement for rockets and then satellites, space funding would not have been maintained at the levels that saw it through the 1950s, 1960s, 1970s, and now the 1980s. Both forms of development went hand in hand with the defence departments maintaining their grip. Looking, for example, at the Soviet launch programme it is easy to see the connection. The surface-to-surface missiles Sapwood, Sandal, Skean and Scarp have all been used as launch vehicles.

The United States too has applied its military hardware to the civilian programme. During the 1950s the United States built and developed the Thor intermediate-range ballistic missile. Thor had its peaceful uses. It was used with other vehicles to send into orbit many types of satellite. At a later date it was matched with the Vanguard rocket stage, renamed Thor-Able and used to launch America's first deep space probe, Pioneer 1. Other rockets were to follow the Thor programme. A modified version of Atlas (the original design for an American ICBM) was used to launch America's first deep space probe, Pioneer 1, and by the civilian element of the US space programme, NASA. By the summer of 1960 the United States had started development of its most powerful ICBM, Titan II. It was this missile that was used to launch the Gemini series that started in 1964 and finished in 1966, during which period there were 10 manned flights. Gemini had obvious advantages to the military. The space enthusiasm in Washington was not as great as it had been earlier and this bothered the US Air Force. Somewhat naturally the Air Force did not like the idea of wasting an opportunity. Here was a manned flight with room on board for simple reconnaissance missions. The photographic detail returned from this flight was described at the time as very successful. At the same time Titan III (a modified version of Titan II) was being developed and it is this rocket that has become the basis for launching most US military satellites.

Before looking at some of the satellites used by the two superpowers and the other nations venturing into space, it is worth remembering that the Chinese and French launchers followed the pattern of development of the USSR and the USA—they based their designs on their military missiles. Equally, it should be said that in the cases of Japan and India—the other two nations which have sent up satellites using their own launchers—missiles were not developed first although the Japanese launchers were based on those developed by the USA. Any country developing a purely scientific space programme will be well on the way to producing a military missile capability. Equally, too much may be made of the military connection with projects that have entirely civilian staffs and aims. However, the fact remains that the vast majority of space programmes have a defence input and many of them have no other function but to serve the military (see figure 7).

As some indication of the military involvement, about 100 defence satellites are launched every year (figure 3). The Americans are quick to point out that most of those spacecraft are launched by the USSR; the ratio is around 85 Soviet satellites to 15 American.

33

But this is not necessarily a sign that the USSR has a bigger military space programme. It merely reflects the fact that the US satellites are designed to stay up for longer periods (figure 8).

Nor is it possible to judge the balance between military and civilian programmes by the amount of money spent on them. The Soviet Union, in line with its policy of secrecy, does not publish details of the costs of its programme. Even if it did, there would be some difficulty in analysing any figures. The importance of any financial outlay on any military project cannot be measured unless it can be seen as a percentage of national wealth and spending and as a fraction of overall defence expenditure. Few in the West have ever accepted the rouble figure produced by Moscow as its overall defence budget. Consequently this has led to somewhat spurious assessments from national and international agencies including the CIA's view that the only way to calculate Soviet spending is to add up all the hardware, armed forces and research programmes and then calculate what it would cost to produce the same amount in the United States and finally present that as a more likely figure.

It is therefore almost impossible to accurately assess the importance in financial terms of the Soviet space programme. The American system of open accounts makes life somewhat easier for the analyst. But pitfalls remain. Recently, the Defense Department's request for funds for space programmes has been higher than the civilian-based NASA request (see figure 7). But picking any one year's figures does not satisfactorily indicate the importance a programme has in a defence department's planning. For example, NASA's civilian programmes tend to have added costs because they are more complicated, especially as it is NASA that directs manned flight projects. Manned flights require more costly procedures, more in-built safety factors, and bigger and more complex launch and recovery processes. None of these comparisons considers the added costs of differing accounting procedures nor the possibility that even if there were no space programmes some of the funds would be requested for other projects. So, weeding out the dollars and roubles spent by the two superpoweres on space programmes is not an entirely practical or reasonable way of illustrating the importance of them to both defence ministries. The simplest way is to look at the types and the roles of the satellites launched during the past few years; this examination clearly shows the importance, the value and, to some extent, future trends.

Military satellites, or satellites with some military value, fall into five broad categories: reconnaissance, weather, geodesy, com-

34

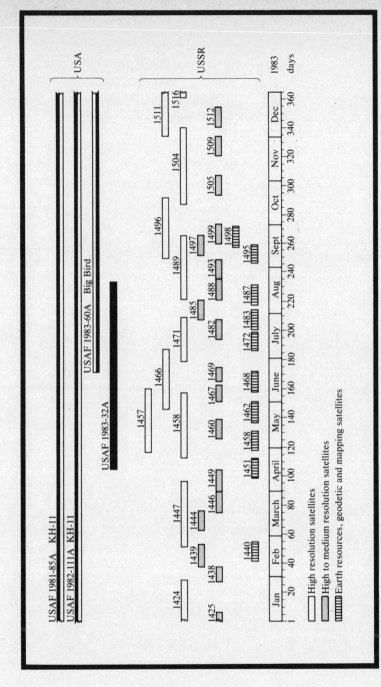

Figure 8. Not a single day goes by without satellite 'eyes' watching the Earth. The figure shows the coverage by US and Soviet photo-reconnaissance satellites during 1983. Each block represents a satellite; the beginning of each block corresponds to the date on which the satellite was launched and the length of the block corresponds to the lifetime of the satellite in days. Numbers above each block for the Soviet satellites indicate the Cosmos satellite number.

munications and navigation. These meet obvious requirements for any military commander: he wants to know enemy positions and strengths; something about the environment including the weather conditions; he needs maps; he wants to be able to communicate with subordinates and formations wherever they are; and he wants his units to be able to pin-point their own positions so as to plan any engagement with the maximum effect.

The advances in space-based meteorology, communications and navigation have been steady but not spectacular. The biggest changes perhaps have been in the different forms of the reconnaissance systems. In the early days these suffered from all the imaginable problems, for example, of being able to 'see' through cloud, of stormy seas and of getting back the information. Some of these problems remain, but they have been eased for the commander by the introduction of new technology and techniques.

Reconnaissance satellites have to be the eyes and sometimes the ears of the modern commander. They photograph areas of special interest—naval dockyards, airfields and troop formations. They eavesdrop on signals and can even loiter in the sky watching for any signs of an attack. And they are plentiful.

Since 1957, more than 2 100 military satellites have been launched. About 40 per cent of them have been used to take photographs. At one time it was necessary to put up two different types of unit to carry out the fundamental *photoreconnaissance* tasks— general and detailed photography. But although this is still sometimes necessary, one satellite is now able to perform both jobs in one mission.

In its basic form, photoreconnaissance starts with a general sweep of a country or region. To do this, the satellite is fitted with a wide-angle lens camera. Once the original sweep, or area surveillance, is checked, points of uncertainty or of new interest may show up. The satellite, or another perhaps, then picks out one particular area or areas and photographs them using a camera with a narrow field of vision and high-resolution film. These satellites operate at about 180 km up, in an orbit which may take them as high as 530 km.

The USSR appears to be still using the two-satellite system to photograph a big area with one and then go for detailed information with the second. The USA has a new generation of what it calls Big Bird satellites which have largely removed the need for this area surveillance. The latest Big Bird—operated by the Air Force—does both jobs. The area search pictures are actually developed on board

36

the satellite and then converted into signals which are relayed to an Earth station where they are reconstituted. The detailed pictures are sent back in film form.

For some years, the Americans have used a 'throw and catch' method. The film is ejected in a capsule from the spacecraft and is caught below by wires trailing from an aircraft (figure 9). This sounds spectacular and the pilots of the aircraft (usually a Hercules) will tell you that it takes "a little time and practice". The capsule makes its final descent by parachute, usually over water, and the whole operation is closely monitored with a back-up ship on standby should the aircraft miss its catch. The importance of these films should not be underestimated, as the information on them is usually very sharp and revealing, although there are draw-backs (see chapter 4).

As we have said, the US satellites stay up for long spells, unlike the Soviet units which remain in space for only a few days. For example, when the first US Big Bird satellite was launched success-fully during 1971 it stayed up for 52 days. Today's version can keep going for anything up to 300 days (see figure 8). One reason for the extended life of these American craft is the advances in micro-technology. This has led to smaller onboard systems which means that more fuel may be carried and therefore the craft is able to remain in space for longer periods. Moreover, the US sensor technology has advanced considerably so that the satellite can orbit at higher altitudes and yet be able to see with the same detail.

It may be that the Soviet programme has suffered from not having the micro-technology necessary to make the weight savings for extra fuel. However, there are indications that the Soviet design and preference is for a satellite that is recovered after about 13 days. Once signal and computer conversion of onboard photography is more readily available, especially in the Soviet Union, there seems no good reason to go to the effort and expense of having to bring back a capsule (see chapter 4).

The Big Bird series of satellites, one of which is thought to have detected the first SS-20 missiles in the Soviet Union, is 'owned' by the US Air Force. But the latest American system of reconnaissance is run by the Central Intelligence Agency (CIA). It is called the KH-11. It is in a higher orbit than the Air Force Big Birds and has one great advantage: it can relay what it sees, when it sees it, back to a ground station. This information comes in the form of digital transmission in what is known as real-time data (i.e., as the satellite

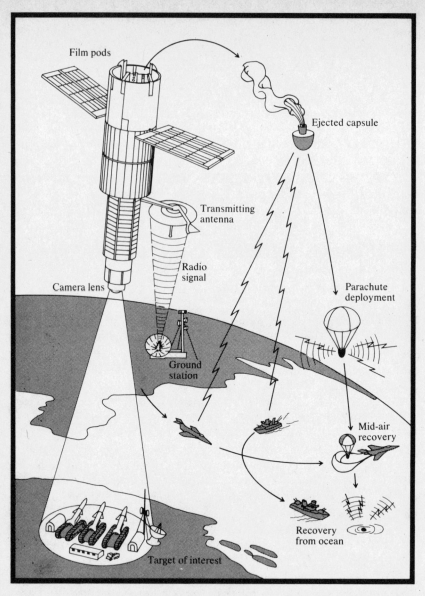

Film pods

Ejected capsule

Transmitting
antenna

Radio
signal

Camera lens

Parachute
deployment

Ground
station

Mid-air
recovery

Recovery
from ocean

Target of interest

Figure 9. Photographic reconnaissance from space often involves a dramatic sequel closer to the Earth. Photographs taken of areas of military interest are ejected from the satellite in a capsule and recovered, as shown here, in mid-air or from the ocean.

receives the information it is automatically transmitted to the receiving station on Earth) and produces an initial computer picture almost instantaneously.

The Soviet photoreconnaissance system is not considered to be so sophisticated, although advances in Soviet miniaturization and their imports by one route or another of Western technology are understood to have improved reconnaissance capabilities. (Soviet photoreconnaissance satellites are contained in the blanket description Cosmos—nearly all Soviet satellites are called Cosmos.)

It should not be forgotten that other nations have interests in photoreconnaissance from satellites. It would seem that the Chinese launched their first photoreconnaissance craft on July 26, 1975. During the same year and again in 1983 two satellites were launched and successfully ejected film capsules. France too has been working for some time towards an independent system. The French were the first after the Soviet Union and the United States to launch their own satellites using their own launching system (see plate 2). During 1981 the French Ariane rocket was tested successfully and this was carried out in conjunction with a $340 million programme to develop a military reconnaissance satellite, SAMRO (Satellite Militaire de Reconnaissance Optique) based on the civilian craft SPOT (Système Probatoire d'observation de la Terre). And a fifth country, Japan, is well on the way to having the ability to launch a similar craft, perhaps with US co-operation, by the end of the 1980s.

So much for the eyes of the space commander. As will be seen in chapter 4, quite a lot is known about the operations and effectiveness of photoreconnaissance satellites. The next grouping is a little more obscure. It comes under the general title of ELINT, Electronic Intelligence.

Electronic reconnaissance satellites have a most important function: to be the commander's ears in space. They must be able to intercept and monitor enemy signals. These signals are generally assumed to be military but diplomatic 'traffic' is picked up and there is no reason to doubt that, as with photoreconnaissance, these systems are used to spy on domestic targets, i.e., the user's own country. Neither of the superpowers gives details of these craft, but it is possible to deduce basic facts. For example, the Soviets tend to put theirs into a slightly higher orbit than the Americans. (Usually the US ELINT satellite is launched from the Big Bird photographic reconnaissance satellite.) Cosmos 1315 is probably an ELINT satellite. It was launched on October 13, 1981, from Plesetsk, about

200 km south of Archangel. It weighs 2 500 kg and has an even orbit; its highest point, its apogee, is 660 km and its lowest point, its perigee, is 620 km. At the time of writing it is still in orbit, as are the two Soviet ELINT satellites Cosmos 1154 and Cosmos 1184 launched the previous year. The latest US system, known as Rhyolite, incorporates important technology for monitoring missile-test signals and could be used in other areas including monitoring at sea.

A sub-group of the Soviet ELINT satellites is used for *ocean surveillance*. It is just as important for the naval commander to be able to detect ships and submarines as it is for the land commander to know where air forces and divisions are placed, in fact more so. The naval commander has the advantage that he can move his forces with less chance of detection than, say, an army commander, who has little or no chance of hiding an armoured division or corps in the middle of Europe. Armies are slow moving, cumbersome and obvious. Ships are none of these things. If the naval commander has this advantage, so too has he the disadvantage of trying to find his enemy.

This was quite clear during the 1982 war between Argentina and the United Kingdom when the Royal Navy relied to a great extent on being able to hide in the huge south Atlantic but had an immediate problem of keeping track of Argentine ships and submarines. It was only when the Argentine Navy returned to port that the job of the British Task Force commander was eased. According to sources in the United States, Washington was supplying the British with considerable satellite cover (see figure 10). Those same sources suggest that the Soviet Union was *not* doing the same for the Argentines although the Soviet photoreconnaissance satellites appeared to have covered the conflict (see figure 11).

The USA started its programme for tracking surface ships in April 1976 with a project called White Cloud. These satellites were known as EORSATs. EORSAT stands for Electronic Intelligence Ocean Reconnaissance Satellite. These satellites carry small sub-satellites to perform electronic surveillance as well as to determine

Figure 10 (opposite). A superpower watches the start of a conflict from space. Here we see the ground tracks over Argentina and the Falkland/Malvinas Islands of US satellites 1980-10A and 1981-85A in March–April 1982 (the number and date of each orbit are indicated on the figure). Intelligence was probably passed on to the UK. By March 27, troop concentration at Rio Gallegos was completed.

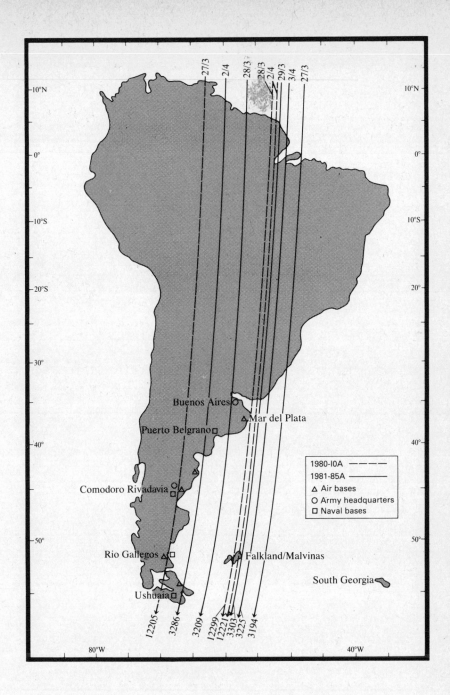

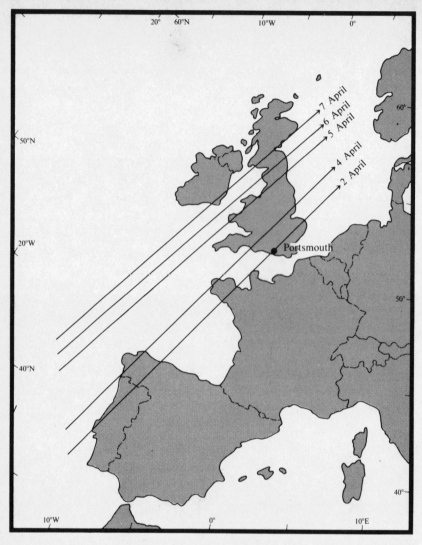

Figure 11. The other superpower watches the British Task Force set off for the Falklands/Malvinas. Here are seen the ground tracks over the UK of the Soviet satellite Cosmos 1347 in April 1982 (the date of each orbit is indicated on the figure).

42

the direction of travel and speed of ships. The Soviet Union operates
such satellites in conjunction with RORSATs, or Radar Ocean
Reconnaissance Satellites; at least they are believed to. Both the
Soviets and the Americans have had problems designing a system
that can search, track and identify ships at sea. As will be seen in
chapter 4 this is not an easy process.

The Soviets are deeply commited to the idea of ocean surveillance
using pairs of satellites. Their first such pair was launched from
Tyuratam in May 1974. Cosmos 651 and Cosmos 654 were sent up
with a launcher based on the Scarp intercontinental ballistic missile.
The satellites are fitted with radar which uses a small nuclear reactor
as its power supply. It was such a system in Cosmos 954, another
ocean surveillance satellite, that caused so much concern when it

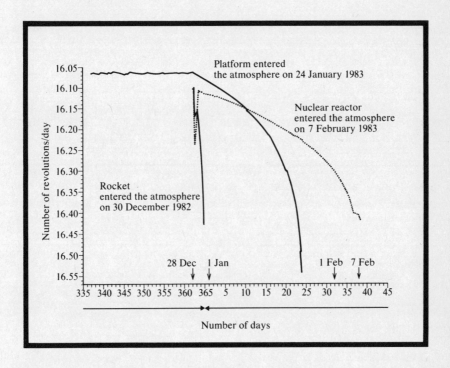

*Figure 12. The story of the Cosmos 1402 nuclear accident can be gleaned from this
chart. On December 28, 1982, the satellite separated from the rocket and the nuclear
reactor. Two days later the rocket plunged into the atmosphere. The world held its
breath until February 7, 1983, when the nuclear reactor, which could no longer be
controlled, plunged into the atmosphere and burnt up.*

43

went out of control and disintegrated over Canada during 1978. It is said to have spread radioactive waste from the nuclear reactor over a wide area of land. A similar accident occurred in 1983 when Cosmos 1402 plunged into the atmosphere, disintegrating with its nuclear reactor (figure 12).

High above these reconnaissance satellites orbit five further types: early warning, nuclear explosion detectors, weather, communications and navigation.

The USA has launched 12 *nuclear explosion detectors*, the Vela satellites, orbiting at 110 000 km. Only two are operational. It is difficult to determine which of the Soviet satellites perform this mission.

Both the superpowers have deployed *early warning satellites* designed (with varying degrees of effectiveness) to spot as quickly as possible signs of nuclear missile attack (see figure 13). The infa-red sensors on board are sensitive to the hot plumes of rockets. Radars used to give 15 minutes warning of an attack but the early warning satellites have extended this time to some 30 minutes.

The Americans have launched a new infra-red test sensor under the project called Teal Ruby. This was sent into space not from a ground-launched rocket, but ejected from the space shuttle during 1983, and the Teal Ruby infra-red telescope is expected to have the further advantage of being designed to detect and track aircraft over foreign territory.

The USA launched the first of a new type of early warning satellite—the Rhyolite—in 1973. It differs from the normal early warning satellite in that it also carries an electronic sensor to monitor missile telemetry during tests. Like all US early warning satellites, this uses an orbit in which it moves round the Earth in the same time the Earth rotates once—24 hours—so the satellite is always over the same point on the Earth's surface. This is called a geostationary orbit, and it allows constant monitoring from its position 36 000 km up. From this vantage point, Rhyolites have watched Chinese and Soviet missile and rocket tests, including the results over the re-entry area in the Far East—the Soviet Kamchatka Peninsula. Theoretically, the latest equipment would allow the Americans to get warning of attack less than two minutes after a rocket launch. In practical wartime terms this would mean that the land-based radars would have confirmation of launch at the earliest possible moment, although even so this would leave only 28 minutes before a land-based ICBM reached the US homeland and much less time for a close-in submarine-launched ballistic missile (SLBM).

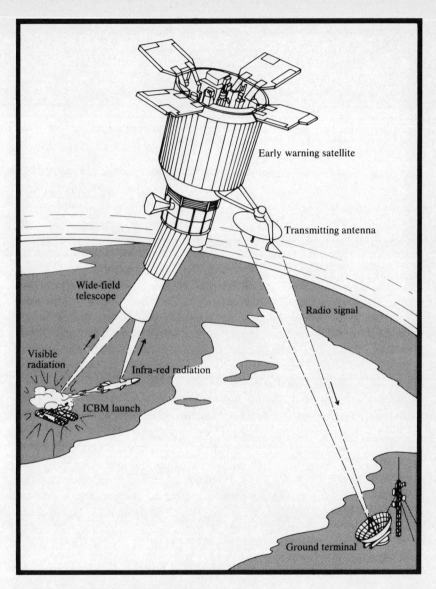

Figure 13. Thirty minutes warning of a ballistic missile attack is now possible due to the early warning satellites. The infra-red eyes of the satellite from 36 000 km pick up the launch of ballistic missiles and the commanders on the ground are informed.

45

The Soviet Union did not find it so easy, or perhaps necessary, to develop the techniques needed to station a satellite over one area all the time. Their first geostationary satellite was probably Cosmos 775, which is said to be monitoring for SLBM attacks. Much of the remainder of the Soviet early warning programme is thought to be carried out by instruments aboard satellites designed primarily for other roles, e.g., the Molniya communications satellites.

It is the *communications satellite* systems that have played a more obvious role in most people's lives compared with other systems. Instant relay of long-distance television pictures is the obvious example.

Such service may not be particularly important to most people, although serious issues have been raised on how, for example, satellites might beam 'unwanted' television into other countries in order to spread propaganda. To the military commander, however, the need to communicate with all his forces is essential. Eighty per cent of US military communications are now carried by satellites.

As we have seen, the Soviets have for some time used the Molniya series of communications satellites. An example is Molniya 3-15, the fifteenth in the Mark III series. It was launched on March 24, 1981 and it remains in a highly elliptical orbit. Its highest point is 39 570 km, whereas its lowest point is only 780 km above the Earth. There have been three series and there has been speculation that some of the craft have a purely military role, though they may also be dual-purpose.

The US military communications system is readily definable. There are three broad projects covering every requirement of the Department of Defense. The first looks after basic command, control and communications under the Air Force Satellite Communications System AFSATCOM, which actually develops and buys equipment needed for certain Defense Department satellites. This does not mean that it is simply a buying agency. For example, AFSATCOM controls multi-purpose communication satellites under SDS, the Satellite Data System.

The second project covers the Defense Satellite Communications Systems (DSCS). Under this are the satellites used by the US President and his chiefs of staff and major commanders in the national and top overseas command headquarters—including the flying command posts such as AWACS aircraft. It is an extremely important and little understood system called WWMCCS—World Wide Military Command and Control System. Linked to it is the communications system for the American intelligence community.

46

Apart from relaying communications some of the satellites are used to pass on information picked up by photographic and electronic intelligence spacecraft.

Thirdly, there is the system that links mobile forces such as ships and aircraft. Most of this work is done by the Navy's Fleet Satellite Communications System (FLTSATCOM), but this is to be phased out and it is expected that the job will be done by AFSATCOM. This does not mean that the Navy has to go elsewhere for *all* its systems, the most important of which is for *navigation*.

The Americans launched the TRANSIT series of satellites to communicate with the early nuclear-powered Polaris ballistic missile-firing submarines. The first successful TRANSIT was launched as a test in 1960. By 1963 the Americans were ready to launch an operational satellite for the Polaris fleet of submarines to obtain accurate fixes. TRANSIT is really an artificial constellation of between four and six satellites orbiting the Earth once every 107 minutes at more than 1 000 km. The constellation is linked to four ground stations updating each satellite's information. The result is a world-wide coverage. Eventually it will be replaced when a much wider-based system called NAVSTAR Global Positioning System (GPS), consisting of 18 satellites, is in complete operation. It is expected to be in total use before the end of this decade and users will be able to fix their position three-dimensionally to within a few metres. The importance of GPS is that it is designed to be used by every element of the US military that needs to know exactly where it or he is. GPS users will range from an infantry soldier with a portable satellite receiver pack to an aircraft carrier, a submarine, a strategic bomber, an intercontinental ballistic missile, or a cruise missile updating its course on the way to its target. The complete system will contribute to unprecedented weapon accuracy (more will be said about accuracy later when we discuss the contribution of two other satellites).

Of the Soviet system not a great deal is known, although much may be surmised from frequencies picked up from some craft. There has been only one official navigation satellite—Cosmos 1000 in 1978. It has been assumed that three systems were needed by the Soviet merchant and military navies to get universal coverage. As expected, the Soviet Union uses satellites in blocks although it is by no means clear that they provide the same standard of accuracy established by existing and developing American systems.

These brief descriptions of the various systems present perhaps an impression of awesome resources, almost as if the whole military

47

system of the superpowers and the developing military nations might be controlled and even safeguarded by the latest generation of space technology. But is this so? Does it work? And, how vulnerable is it?

4. The military need

At a time when technology is presented as so wonderful a thing that it is said to have an answer for every need, it is easy to believe that satellites hold the keys to international security. Many accept that technology advances at such speed that any claim presented in a manner of some authority is most likely to be true. In short, we have come full circle through the centuries from the times when our forefathers would sit and gape in amazement and belief as travellers told of exotic sights in far-off lands, and because our ancestors wished to believe, they did. So it is with reports of satellite technology and its capabilities. The general public has come to believe that satellites can indeed tell the rank of an infantry soldier or overhear a telephone conversation between two executives or ministers travelling at speed in their limousines. And the escape clause to this naivety is that if it is not so now, then it soon will be.

Satellite systems designated for various military roles have not always been able to fulfil them. For example, for many years both the Soviet Union and the United States have recognized the importance of ocean reconnaissance. The need to pin-point fleets and flotillas and then to monitor their operations and signals has been a major priority for both superpowers. Satellites to do this job have been in orbit for many years, but they have not had great success. Technically, the projects have always seemed feasible enough, but many 'users' have not been satisfied with the results. Satellite sensors have often found it difficult to penetrate even moderately rough seas such as a force seven wind may whip up, so that ships have been able to 'hide' from satellites in adverse weather conditions. But in general terms the ability of space-based systems to supply commanders with intelligence is improving so quickly that, in the West at least, the only restriction on rapid improvements is likely to be the priority that any one area receives rather than the technical limitations.

The need to get an effective reconnaissance space programme

49

started was never more evident than when the Soviet Union shot down Gary Powers in 1960. Powers was overflying the Soviet Union in his U-2; his mission was to collect data on the rocket launch area at Baykonur. It should be remembered that the Powers flight took place at a time when satellite reconnaissance was in its infancy. Sputnik 1 had been launched only three years before and it was not until 1960 that the Americans made their first recovery of a capsule (and therefore the potential to bring back photographic intelligence) from a spacecraft—Discoverer 13. However, the incident did concentrate the efforts of the United States, at least, to produce a comprehensive space-based intelligence gathering system; and in spite of setbacks, the programme advanced with tremendous speed. The following year, 1961, the United States launched SAMOS (Satellite and Missile Observation System) and with it the basis of their spy-in-space systems.

Pictures produced from satellites are so commonplace that it is often assumed that this is a relatively easy form of intelligence gathering. In the correct conditions this may be so. In spite of extraordinary advances, photoreconnaissance satellites are reliable only during daylight and good weather conditions. But during those ideal conditions the results are very impressive. A photoreconnaissance satellite is capable of picking out objects measuring not much more than 30 cm—and this from a height of something like 200 km. And the more contrasting the surroundings to the object, the more definite is the identification.

A simple form of satellite camera feeds the film through two lens units—one with a short, and the other with a long focal length. As the film is used, it is pulled across a plate containing a combined developer and a fixer, to stop over-developing. The negative is dried by an automatic heater and then rolled on to a spool for electronic scanning. This is simpler than it may sound. A narrow beam of light runs in horizontal lines across the negative. The light passes through the negative but, depending on what is on the film, at different intensities. Behind the film is a signal generator. As the beam penetrates the film it is 'read' by the generator which sends out a signal matching the intensity of the beam. The series of signals transmitted to Earth stations can be converted back into pictures by replaying the intensity of the signals over a plate. In other words, the beam scans the negative and the strength of that scan is relayed in electronic signal form and reproduced on the ground. This system, developed during the early days of photoreconnaissance, worked well when it was used in conjunction with detailed photographs received when the satellite ejected capsules of film. The Americans recovered these in mid-air by a specially converted transport plane or when the capsule had splashed down, while the Soviet capsules were collected after a soft parachute landing.

An important advance came with the introduction of digital data and false colour photography. Digital data are produced by a scanner which uses an array of small, light sensitive sensors that electronically convert the images into digits. Scanners can operate in several different colour bands plus an infra-red band. The data are transmitted in digital form back to a ground receiver station and then the digital image is rebuilt.

This method allows the ground users to give their own colours to different parts of the picture. The data enables them to distinguish between different types of material even if they are the same colour. So, if a painted surface (which is recognized by its special combination of colour bands) is given one colour in the photograph, say blue, and foliage another, say red, then it is possible to separate a green painted tank camouflaged to blend with surrounding green vegetation. The green tank shows up blue and the real foliage, red. Furthermore, using digital photography, it is possible to pull out parts of a picture that could easily be lost in conventional photography. It may be nothing more than the manipulation of data by the computer. It may be the computer and not the human eye that can rebuild part of a picture lost due to distortion and in doing so reveal intelligence that would never have been found by traditional methods of photo-interpretation (see Plate 3). The combination of 'false-colour' photography incorporating infra-red regions of the spectrum (known as multispectral photography) and the use of computers to extract the information, both apparent and hidden, is the way ahead for military photoreconnaissance for the rest of this decade.

The commander is gaining a huge advantage over the earlier systems: not only is the information more precise, but it can be transmitted in real time. Hence, the commanders are getting a new weapon in the capability to extract from space 'moving' intelligence—information about the enemy or the environment almost as events happen.

There have been advances in the technology needed for photoreconnaissance to overcome darkness and bad weather. The most obvious way through poor conditions is the use of radar and signals intelligence. The USSR took the lead in the development of space-based radar and have since kept it. One significant development has been the introduction of Synthetic Aperture Radar, often abbreviated to SAR.

Rather like a torch beam, a conventional radar scan at great distances is V-shaped with the highest intensity of the radar's beam at the centre of the 'pool of light'. Consequently the edges of the radar picture are not so clear. It was known that this problem could be overcome with a longer antenna. But that presented obvious problems because the size of antenna that can be used is to some extent governed by the size of the satellite and the weight and size of the antenna carried into space. (The space shuttle could be used to overcome this by taking up an antenna in sections and assembling it in space.)

Synthetic aperture radar does nothing more than give the *impression* of having a huge antenna. As the antenna (equivalent to the aperture of a camera) moves along with the satellite in its orbit, the radar signals reflected back from the objects on the Earth's surface are received by the antenna and recorded. They are then later combined or synthesized. The maximum length of the synthetic aperture is then the length of the satellite path along which the moving antenna receives the reflected signals from a particular target. The result is a series of layered views of the object of equal intensity throughout the scan and therefore almost perfect radar. Just how good and important this is may be seen in the American Seasat images (see plate 5). Using synthetic aperture radar it was able to monitor clearly 100 km swathes. The development of a radar with a synthetic aperture allowed the radar to look from different angles and therefore get a more constant view of the target area. And the result has been radar pictures that are nearly as good as ordinary high altitude photography. This would obviously give an immense advantage to a commander who can get back some form of picture even if there is cloud cover at night.

But there remains another problem. These radars need big energy sources—powerpacks. The Soviet Union decided that the simplest way to do this was to fit the satellite with a nuclear reactor. They launched their first system, used for ocean surveillance, in 1974. One of the easiest ways of allowing a satellite to detect and monitor a target is for that target to shine a 'torch' at it, in effect saying: 'Here-I-am'. The 'torch' is a signal, such as a radar beam. So some satellites spend time 'looking' to see whether ships switch on their radars or signal other units or shore bases. The satellite picks up the radar or radio beams and monitors.

These spacecraft are known as ferrets and they are not restricted to ferreting out ship's radar. They can, in theory, be used for any task where electronic intelligence is needed by commanders in peace-time as well as wartime. One obvious role is that of the American Rhyolite satellite which has been used to pick up signals from Soviet missile test flights.

This is not simply a matter of the USA checking on the latest Soviet tests in order that they might be compared with American systems and targeting plans. As the development and testing of rockets is an integral part of some arms control agreements, it is important that these types of satellites are able to monitor the activities of either superpower. Indeed the various two-power (bilateral) arms control treaties tend to protect these satellites for this very reason. Such treaties include a promise not to interfere with the 'national technical means of verification', a recognized euphemism for reconnaissance satellites. Of course, there is no suggestion that satellite monitoring will stop breaches of agreements, but the knowledge that if you cheat you will probably be found out may have some effect. In any case, the use of satellites for verification—checking that an agreement is being kept—is built into both sides' negotiating programmes. For example, when the Americans were considering ideas for the MX missile system, a suggestion was made that inspection hatches should be built into the roofing of the missile shelters. The thought was that the Soviet Union could send satellites over that would be able to count the numbers of missiles and therefore this would in some way help overcome the problem of verifying bilateral agreements.

At the same time, during the development programme, tests were carried out to fool the satellites. One idea for MX was that a number of dummy missiles should be built and that these would be moved about on huge trailers so that the Soviet Union would find it difficult to tell which was the real missile and which was the mock-up. In conversation with one of the authors of this book, an MX project engineer demonstrated how the dummy had to be exactly the same weight as the real weapon because it was considered that Soviet satellite observation was so good that it would be possible to record the engine strain and downward pressure of and on the trailer and thereby whether or not the missile was real or a mock-up—a reminder perhaps that satellite monitoring is no guarantor of agreements, nor is it a means of preventing war.

Although satellites have been used by both the Soviet Union and the United States to monitor crisis areas, there is no evidence to suggest that wars have been prevented or stopped by one or both

53

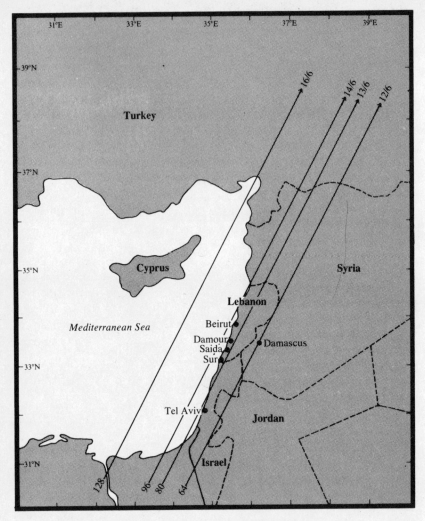

Figure 14. The USSR observes conflict in the Lebanon, June 1982. Ground tracks of the Soviet satellite Cosmos 1377 are shown over the Lebanon (the number and date of each orbit are indicated on the figure).

superpowers knowing troop deployments. An example of this was the Middle East War of 1973 (figure 5) and more recently the invasion of Lebanon in 1982 (see figure 14). War was not prevented although the Soviet satellites may have observed the build-up of the Israeli forces along the border. Soviet satellites certainly observed the conflict between June 12 and June 16, 1982. Another example of satellite reconnaissance *not* preventing war was Operation Corporate, the codename given by the British to the 1982 war with Argentina in the South Atlantic (figures 10 and 11). An American reconnaissance satellite overflew Argentine bases six days *before* the Argentine invasion of the Falklands. On March 18, 1982 and on April 2 (the day of the invasion) the same satellite flew over the islands themselves.

On the day of the invasion, the Soviet Union launched its own reconnaissance satellite, Cosmos 1347. This satellite made two passes over Portsmouth, where the bulk of the British Task Force was based. The first orbit over Portsmouth was on April 2. Two days later, as the main Task Force elements were preparing to put to sea the following morning, Cosmos 1347 made its second flight over the region. There had been a great deal of speculation about the Royal Navy's use of nuclear submarines and on April 5, 6 and 7, Cosmos 1347 orbited over Britain's main nuclear submarine base at Faslane on the west coast of Scotland. Faslane is home also for the Polaris ballistic missile submarine fleet. Cosmos 1347 was a high-resolution reconnaissance satellite which stayed in orbit for 50 days.

This was only one of 35 photoreconnaissance satellites launched by the Soviet Union during that year. The Americans launched two (if this seems an imbalance it should be remembered that US satellites have a much longer life than those launched from the Soviet Union). The photoreconnaissance spacecraft outnumbered all other systems during that year—as they do normally. The Americans did launch an ELINT (electronic intelligence) satellite athough it was not launched from the ground, but ejected from one of the three photoreconnaissance satellites mentioned above. The USSR launched six ELINT systems, Cosmos 1335, 1340, 1345, 1346, 1356 and 1400, and they sent up five early warning satellites for detecting launches of missiles. (The USA launched one.) But for many observers, 1982 was significant for other reasons.

It was in this year that the shuttle carried its first military payload. This was quite important, for there had been a noticeable reluctance in Washington to talk about the military capability of the shuttle. Privately however, the military were very pleased. They pointed out that they would be failing in their duty if they did not take advantage of every new development in space. Anyway, it was only a public relations exercise because the funding and the programme details were published in official and unofficial documents and journals.

The military payload was an experimental sensor for a project mentioned earlier, Teal Ruby. It has been suggested also that this project in the shuttle was another turning point for the whole military space programme. It was as if it started to gather momentum.

Of the planned 72 shuttle flights, 35 per cent are scheduled to have military payloads. One of the payloads will be the test equipment for a project called Talon Gold. Talon Gold is part of what was dubbed President Reagan's 'Star Wars' concept. It is concerned with a low-powered laser to pick out targets, track them and mark

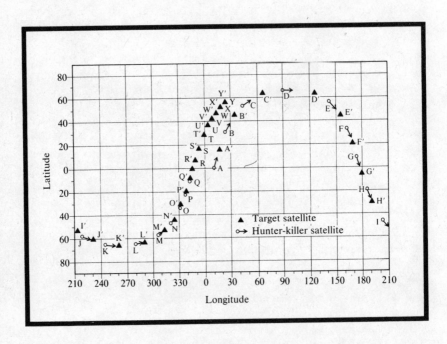

Figure 15. The hunting of target satellite Cosmos 1375 by the interceptor Cosmos 1379. The triangle represents the ground track of target and the dot on the end of the arrow represents the ground track of the hunter-killer satellite. The changing position of the target is indicated by the letters A, B, C, etc. and of the interceptor by A', B', C', etc. The interceptor caught up with the target satellite at V/V'.

Plate 1. The launch pad at the Baykonur rocket base. From such a pad the spacecraft Vostok 1 was sent into orbit around our planet with the world's first spaceman on board—Yuri Gagarin.

Plate 2. An Ariane rocket taking off from the French launch site in French Guyana, a reminder that the space race is not just between the two superpowers.

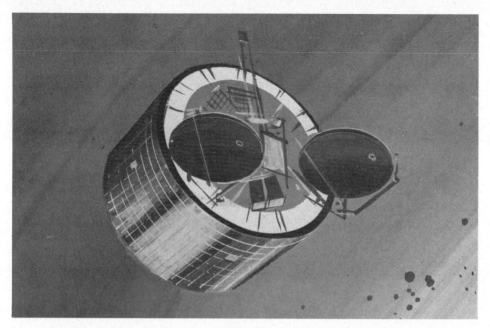

Plate 3. Modern military forces have a greater need for very fast and reliable communications. This is provided by military communications satellites, such as the one shown in this artist's impression.

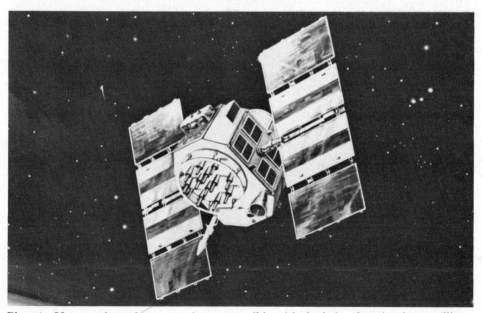

Plate 4. Unprecedented accuracy is now possible with the help of navigation satellites, such as the US NAVSTAR satellite depicted here. To be able to deliver warheads accurately requires exact knowledge of position, velocity and direction of the delivery vehicle. Constellations of navigation satellites provide continuous world-wide capability for any weapon system.

Source: US Air Force.

(a)

Plate 5. With the help of the synthetic aperture radar pictures can be taken through clouds, rain and in darkness. Such pictures are illustrated below and opposite: (a) is an image of Los Angeles, California, taken from a US satellite orbiting at 700 km at 8.30 p.m. local time—details such as small pleasure boats, roads and railways can be seen; (b) Cologne, FR Germany, showing the city airport, barges and bridges.

Source: Macdonald Dettwiler and Associates Ltd. Canada.

(b)

Source: Processed by DFVLR, FR Germany.

Plate 6. Computer-enhanced photographs of the Soviet Nikolaiev 444 ship yard in the Black Sea. The photographs were taken from a US military satellite in July 1984.
(a) An oblique view of the dry docks, the two sections of a nuclear aircraft carrier under construction can clearly be seen in the dry docks. Some of the other features include a foundry in the foreground and assembly shops behind.

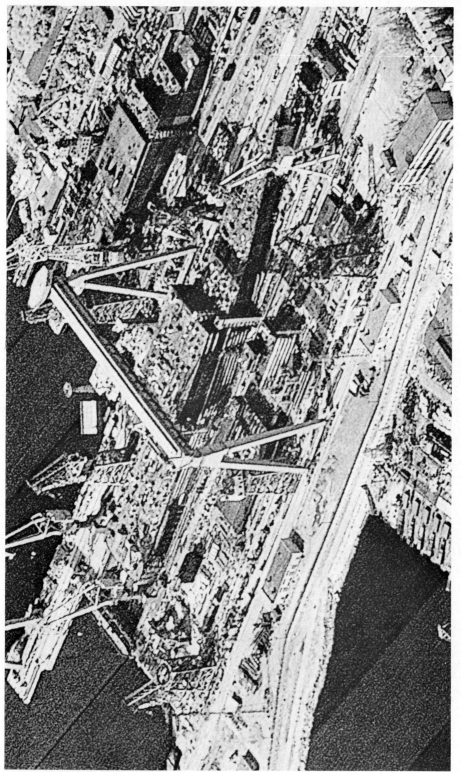

(b) A more close-up view of the Soviet nuclear aircraft carrier.

(*c*) An almost vertical view of the aircraft carrier. The major portion of the ship is 264 m in length and the stern section is 73 m in length.

Source: Jane's Defence Weekly

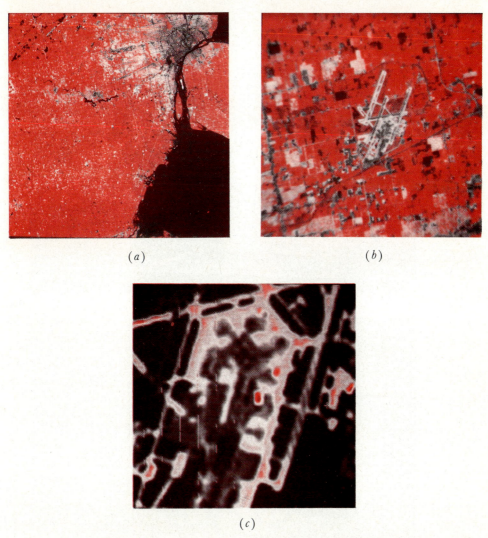

(a)

(b)

(c)

Plate 7. Reconnaissance from space has been revolutionized with aid of the computer and micro-electronics.

These processed images[a] illustrate what can be seen from 700 km altitude by a US civilian satellite with the ingenuity of modern technology. This is three times better than the civilian satellite pictures of the last decade but 50 times inferior to satellite pictures available to the military.

(a) Photographs from 700 km altitude of the Detroit, Michigan area.
(b) An enlargement of the main civilian airport.
(c) This is a further enlargement of the airport building using false colour techniques. The computer was instructed to assign red colour to shiny objects such as aircraft.

[a]For the purposes of this book only two colours have been used.
Source: Orhaug et al., 1983 Landsat 4 TM-Data: examples of resolution capacity, *FOA Report C 30329-E1*.

(a)

(b)

Plate 8. From Sputnik to space shuttle in just 24 years—and what next? Will the space shuttle be used to ferry up parts for a laser battle station or for a military base in space? Or will it be used only for the betterment of mankind?

(a) In this artist's concept the US space shuttle is placing a platform in orbit 36 000 km above our planet, which could provide services for use on Earth: telephone and TV communications; communications relay between land vehicles, ships and air-craft; observations for meteorology; and scientific investigation.

(b) The USSR is also developing a kind of space shuttle; here the Soviet experimental re-usable spacecraft is being recovered.

Plate 9.   The American flying laser laboratory.   The laser beam is emitted from the circular window on top of the NKC-135
Airborne Laser Laboratory aircraft, a converted Boeing 707.

Source: US Air Force.

Plate 10. A US F-15 aircraft carrying an ASAT missile under its belly.

them for destruction. The test targets will be high-altitude aircraft and perhaps satellites.

With this in mind, and with so much of the East–West military system so obviously relying on spacecraft, it is not surprising that the second feature of 1982 was the Soviet testing of hunter-killer satellites. As already recorded, Cosmos 1379 was launched from Tyuratam on June 18. Its task was to hunt down and kill Cosmos 1375, launched on June 6. It succeeded (figure 15). This is thought to have been the 49th satellite involved in some 20 tests carried out by the Soviet Union since 1963. It was that far back—more than 20 years ago—that both superpowers started to get the idea that space could be the battle-ground of the future.

5. Space as a battlefield

It was inevitable, certainly understandable, that the 'Star Wars' speech of President Reagan in March 1983 should generate misconceptions and excitement. As much as it is possible to abhor adoration of any weapon system, it is human nature that the possibility of such exotic weapons should generate awe, wonder and fantasy. In spite of the illustrations and speculations that followed that March 1983 speech, it is worth repeating that the US President said nothing more than that the possible use of beam weapons and even space-based systems were things of the future and would it not be a fine thing if technology and ingenuity could be combined to provide a defence against intercontinental ballistic missile attack. At no time did he promise that the technology was within easy grasp of scientists. Nor is it.

Although the President's emphasis was on some anti-ballistic missile (ABM) systems, the general public's attention has focused on space battles—that is on anti-satellite (ASAT) warfare; and, of course, there is a connection in spite of the differing functions of the two types of weapon system. The projected ballistic missile defence (BMD) system is vastly more complicated than those for anti-satellite use. However, the basic technologies are so similar that research and development work on BMD will produce an ASAT system as a spin-off. It has only to be imagined that if it is possible to produce a system that will destroy a series of fast-flying ballistic missiles then that same technology slightly modified should have no trouble in dealing with a satellite. This is why few believe that it would be possible to go ahead with an anti-missile system and at the same time ignore the temptation to drain off ASAT systems as by-products.

Of course, anti-satellite weaponry does already exist, but it is in its infancy. Tales of fantastic advances and capabilities should, in the mid-1980s, be treated with caution. No weapon system exists that could knock out the complete range of military satellites used by both superpowers. Reports of tests by the Soviet Union, and more recently by the United States, suggest that anti-satellite warfare is at a crude stage in its development and that neither side is capable of hitting anything but those satellites orbiting quite close to the Earth in the so-called near-Earth orbits.

There is nothing new in the concept of ASAT development. Once the superpowers had got into space, two complications presented themselves immediately: it was necessary to develop a means of destroying the other's satellites, and some way had to be found to protect their own spacecraft. The first part of the problem could be modified inasmuch as it was not necessary to actually destroy an opposition system. After all, the function of military spacecraft is, in general terms, to gather information and relay it to commanders either in the field or at some supreme headquarters. So all that is required of an anti-satellite system is to stop that happening. In other words, the opposition spacecraft does not have to be destroyed, only degraded or 'wounded' by, for example, blinding its sensors or jamming its communications, so that it can no longer function. Consequently the theory might be that the ASAT system does not have to have the degree of sophistication that many believe it should. But as the second part of the two-part problem is resolved, i.e., protection of spacecraft (the military often use the term 'harden' rather than protect), there is an argument that sophistication becomes more necessary. Today, however, there is little evidence to show that 'hardening' has advanced far enough. (In fact, at present, one of the best protections would be altitude because of the relatively limited range of the existing ASAT methods.)

One of the first obvious references to Soviet ASAT weapons appeared in 1962, five years after the first Sputnik was launched. It came from the Soviet leader Nikita Khrushchev. He is said to have claimed that the Soviet Union had developed a very accurate missile that could knock out American systems. In Mr Khrushchev's words it "could hit a fly in outer space". There is no hard evidence to suggest that Mr Khrushchev's claim was backed by any technology. Western experts—following the military tendency to assume the worst of the enemy—have generally accepted what he said as being true. But the early 1960s were particularly difficult for the Soviet Union in general and its leader in particular. At the time, "Krushchev's constant boasting had become a joke, and his reckless plunging and endless expedients were producing a widespread mood of frustration and unsettlement" (*Krushchev* by Edward Crankshaw).

Certainly a great deal of effort had gone into the development of Soviet rocketry, so much so that the conventional forces had suffered and there was at the time in Moscow at least a temptation by some senior officials to boost claims of the Soviet missile capability in order to justify what was an enormously resource-consuming programme for the Soviet Union and its military.

Nevertheless Khrushchev's boast was thought by some analysts to be perfectly realistic because the expensive Soviet rocket programme had included a ring of rocket sentries around Moscow. These were the Galosh missiles. Galosh is the Western name given to the Soviet Union's missiles for anti-ballistic missile use. They are based around Moscow and their role in wartime would be to destroy incoming Western missiles aimed at the Soviet capital while those missiles were quite close to the target area. Each rocket is about 20 metres long and is thought to have a range of something in the order of 300 kilometres, which would not give it much leeway if it were aimed at a satellite; but more than 20 years ago the Khrushchev claims plus knowledge of the Galosh system aggravated the suspicions of American defence planners. Not only did this turn the minds of some military engineers towards developing further methods of anti-ballistic defence, it also influenced the decision that ASAT engineers could advance more quickly if they concentrated on Earth-based systems rather than on space-based systems. The Soviet Union, however, did not move in this direction.

Whatever the claims of Nikita Khrushchev, Soviet scientists knew that they did not have to be able to hit a fly in space to develop a reasonable ASAT weapon. They did not need to hit anything. The requirement was to do no more than stop an alien satellite from functioning.

So in 1967 the Soviet Union started a series of tests to do just that. A satellite is sent into orbit quite close to Earth, say a couple of hundred kilometres up. Shortly afterwards, perhaps two or three days later, another satellite is launched; this is the interceptor (see figure 16). When the interceptor has manoeuvred quite close to the target, it 'blows itself up' using a conventional chemical explosive. The Soviet Union appears satisfied that the principle of hunting and killing without actually destroying the target is all that is needed. However, the tests have varied considerably during the experimental programme which has been going on now for more than 20 years. At the time of writing, the Soviet Union has carried out some 20 tests. The first batch was between 1968 and 1971 (see table 2); the second between 1976 and 1978. Since 1980 there have been

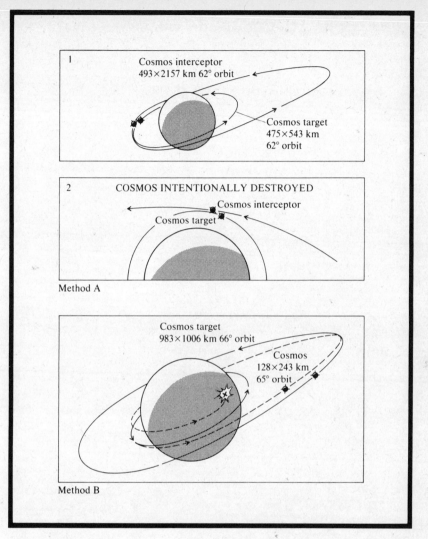

Figure 16. The Soviet hunter-killer system in action. This can destroy satellites as far away as 2 000 km but is a ponderous system since it can only easily hit satellites in certain orbits. Three kinds of test are carried out. In A the hunter killer approaches its prey when it is closest to the Earth and then explodes. In B the hunter-killer intercepts the target at the furthest point from the Earth and is recovered. In the third method (not illustrated here) the two satellites are not orbited in the same plane.

61

Table 2. The Soviet interceptor-destructor satellite programme tests 1968–83.

	'Target' satellite		Associated 'interceptor' satellite(s)	
No.	Cosmos no.	Launch date	Cosmos no.	Launch date
1	248	19 Oct 1968	249	20 Oct 1968
2			252	1 Nov 1968
3	373	20 Oct 1970	374	23 Oct 1970
4			375	30 Oct 1970
5	394	9 Feb 1971	397	25 Feb 1971
6	400	18 Mar 1971	404	4 Apr 1971
7	459	29 Nov 1971	462	3 Dec 1971
8	803	12 Feb 1976	804	16 Feb 1976
9			814	13 Apr 1976
10	839	9 Jul 1976	843	21 Jul 1976
11	880	9 Dec 1976	886	27 Dec 1976
12	909	19 May 1977	910	23 May 1977
13			918	17 Jun 1977
14	959	21 Oct 1977	961	26 Oct 1977
15	967	13 Dec 1977	970	21 Dec 1977
16			1009	19 May 1978
17	1171	3 Apr 1980	1174	18 Apr 1980
18	1241	21 Jan 1981	1243	2 Feb 1981
19			1258	14 Mar 1981
20	1375	6 Jun 1982	1379	18 Jun 1982

Other Cosmos possibly related to interceptor-destructor programme are nos. 185, 217, 291, 516, 520, 521, 752, 816, 844, 885, 891, 933, 1075, 1146 and 1169.

Source: Jasani, B., 1984, Outer space—battlefield of the future, *Futures*, **14**(3), 660.

spasmodic operations the most ominous of which, in some Western eyes, took place during June 1982. This was the one in which ASAT was used as part of an overall nuclear war exercise.

On June 6, 1982, the Soviet Union launched a satellite described by the official news agency TASS as Cosmos 1375. Cosmos 1375 was orbiting at between 990 and 1030 km above the Earth and was taking about 1 hour 45 minutes to complete an orbit. The hunter-killer was launched 12 days later from Tyuratam on June 18 shortly before 1500 hours. This was Cosmos 1379. The hunter-killer was going round the Earth faster than its target and at a different angle to the equatorial plane (a different 'orbital inclination'). Ground control altered the inclination and the speed of the hunter-killer while it was still in its first orbit. The next time round Cosmos 1379 intercepted and 'killed' Cosmos 1375.

Soon after launching the ASAT satellite, the Soviet Union launched two intercontinental ballistic missiles, a submarine-launched ballistic missile and an SS-20 medium-range missile. The ABM system was then activated. If these weapons were co-ordinated with the ASAT test (as Western intelligence has suggested) then they provided, for the first time, an actual nuclear war scenario; a play-out of a nuclear war started by the destruction of a military satellite. The reminder that war may start in space occurred, ironically, on the eve of the United Nations Second Special Session on Disarmament and a few weeks before the second UN conference on peaceful uses of outer space.

It is difficult to determine how successful the Soviet tests have been. So far it would seem that the interceptor satellites have managed to manoeuvre quite close to the prey. There have been reports of the interceptors self-destructing, which is not a very surprising indicator that the Soviet Union does have the technology to remotely explode the hunter. It would appear, and this is open to speculation, that the Soviet scientists have not gone so far as to destroy the target satellite. This would seem reasonable rather than a question mark over their ability to do so; the experiment may be satisfied by demonstrating the capability rather than going to the extreme of actual destruction. A variety of standards have been put forward by American analysts, including the closeness of the hunter-killer craft, which suggest that the Soviet Union has a success rate of something in the order of 65 per cent. This is considered adequate at this stage of their programme. Added to this should be the range of these tests. The highest altitude reached in a Soviet ASAT test is approximately 2 300 km, within the range used by US recon-naissance weather and Transit navigation satellites, as well as by the space shuttle. NAVSTAR navigation satellites at 20 000 km and communications and early warning satellites at geosyn-chronous altitude (35 800 km) are out of reach of the current Soviet system.

However, there are indications that the Soviet Union is working on the idea of putting hunter-killers on bigger rockets with more effective boost stages, and if they do so, then some of the higher Western spacecraft will be particularly vulnerable. But at the moment, the Soviet programme appears confined to the basic near-Earth attack system. This is also the case for the US programme.

Some say that the USA's venture into the ASAT business was a direct response to Nikita Khrushchev's 'fly in space' boast. Others have claimed that it was more by chance. For example, Tom Karas in his book *The New High Ground* writes:

63

the first test of an ASAT weapon was an accident. In October 1962, the Atomic Energy Commission and the Air Force carried out a high altitude nuclear weapon test codenamed STARFISH. STARFISH, as it turned out, fatally damaged a number of orbiting satellites. What's more, not one of them was in line of sight of the nuclear detonation. They were all wrecked by the massive doses of high energy electrons that the weapon had suddenly injected into their paths.

This should not imply that the Americans stumbled on the idea of an ASAT programme. The Khrushchev speech was in 1962. In fact, the US ASAT missile weapons date back to 1959 when under a programme called Bold Orion the US Air Force began testing ASAT missiles launched from B-47 aircraft. Only four tests were conducted, but in those days the guidance technology was inadequate. Therefore nuclear weapons were chosen for an ASAT warhead. In 1963 President Kennedy gave the go-ahead for an anti-satellite project. The US Army worked on a programme that involved the Nike-Zeus missile as the carrying vehicle for a nuclear warhead. The idea was simple: launch the Nike-Zeus and then explode the warhead more or less overhead and in the region, but not necessarily in the path, of a Soviet satellite. That programme started in 1963 and finished the following year.

Meanwhile, the Air Force had been working on a system involving the Thor missile. Thor was the name given to an intermediate-range missile whose development started in 1955. It was launched as a weapon system in 1957 (the same year as Sputnik 1) and had a range of some 2 400 km which meant that it was reasonably limited to deployment in Europe. The Air Force started ASAT tests in the Pacific from Johnston Island and in 1964 the Thor ASAT system was considered to be operational and stayed so until 1975 when it was taken out of service. The early experience that Tom Karas describes suggests the immediate and obvious drawback in the Thor system which relied on a nuclear warhead. Firstly, because of the 1963 Partial Test Ban Treaty which included a ban on nuclear tests in the atmosphere and in outer space, the whole system could never be fully tested. A dummy warhead simulated weight during the test flight and the rocket was quite accurate; some of the tests came very close to the targets. But it wasn't the accuracy that was doubted. The problem was that although the nuclear warhead would undoubtedly destroy or degrade an enemy satellite, it would quite probably do the same to American spacecraft. Consequently the

Thor programme with a nuclear warhead would be nothing more than scientific suicide—so it was discontinued.

The current American ASAT system has a great advantage that it does not need a nuclear warhead. Although it is not yet in service, nothing in the planning of tests so far suggests that this is likely to change.

The simplest description of the new American ASAT weapon is a fast jet aircraft carrying an anti-satellite missile (figure 17, plate 10). It would appear that the missile does not carry a warhead that is designed to explode in order to damage the target satellite. Instead, it is what military engineers call a hittile; it is designed to destroy or damage the target by ramming it.

The aircraft is an F-15 interceptor, the same sort of jet that the United States Air Force deploys throughout the world in its tactical

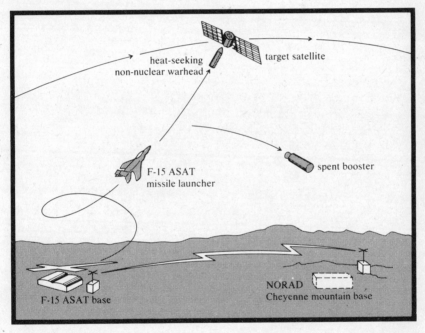

Figure 17. Conventional space warfare, US-style. Here the new air-launched ASAT missile is shown attacking an enemy satellite. The 'warhead' is a non-nuclear miniature homing vehicle guided by an infra-red sensor to its target. The prey is killed by ramming. The system can reach satellites orbiting at about 450 km. Since it can be launched even from civilian airports it has considerable flexibility; and probably by 1990 a relatively swift attack could be launched against a whole range of low orbit satellites.

65

air squadrons. A special carrier attachment, normally called a pylon, has been designed by Boeing Aerospace for the F-15 so that the plane can carry the missile, which is about 5.5 m long and just over 50 cm in diameter. The missile is a package of a two-stage rocket (one stage made by Boeing and the other by Vought) together with the all-important homing device. It is this part of the system, known generally as the MHV—miniature homing vehicle—that would do the damage. The MHV is fitted with small sensors that are designed to pick out recognition points on the target, including heat, so it can home in on the target. The method of operation would be to take the F-15 to a maximum altitude under the orbit of the target satellite and then set the 48 000 km/h rocket on a general course for an intercept point. This course can be corrected by the missile's own system and both the F-15 pilot and the MHV will be under the control of Space Defense Operations Centre inside Cheyenne Mountain, Colorado, using the so-called Space Detection and Tracking System (SPADATS). A test of the ASAT system without its target and without its 'warhead' was carried out on January 21, 1984.

It is thought that the US system could be operational by late 1985 and that some 28 weapons will be ready for use on a squadron of F-15s by 1987. Two squadrons will be fitted for ASAT missions by 1989, but it is expected that the plan may be delayed by as much as two years depending on tests against targets in space.

It would seem that the US system has the same major drawback as the Soviet ASAT weapon in that it cannot reach targets other than those in close or near-Earth orbit. But these do include reconnaissance and navigation satellites and Salyut space stations. There could be stages in the flights of certain Soviet communications satellites when they would be vulnerable. These weak points would tend to be when the satellites were at the lower levels of their flight and these would probably be when the spacecraft were far away from the American mainland.

However, once the Americans have their MHV in operation they will be far ahead of the Soviet Union in ASAT weaponry. The Soviets have the problem that because they have to launch their hunter-killer satellites from rockets, they are confined to certain launch sites in the Soviet Union. It would also seem that because of this the Soviet system is restricted because it has to 'wait' for the target satellite to come into its zone and therefore it could take over a day before a hunter-killer was able to take on a specific target. And if the plan was to destroy a number of satellites this might take days to execute.

The American system, however, being based on a simple aircraft, can go anywhere in the world where there is a friendly air-strip and take off from there. Considering the world-wide interests and facilities of the US Air Force and the enormous treaty understandings for aircraft deployment, it follows that the United States could position its anti-satellite system at short notice anywhere on the globe and would be able to mount a co-ordinated attack at almost any time of its choosing. That, at least, is the hypothesis.

There are other ideas for ASAT weapons that have caught the imagination of strategic planners and even those who see some political and even commercial opportunity.

One suggestion is space mines. These would be planted in space quite close to major satellites and detonated from ground control. Others have thought of having a satellite that 'goes to sleep'. The idea is that it is launched, monitored world-wide for a short spell and then formally pronounced as either malfunctioning or 'dead'. The satellite is left to bide its time until ground control wakes it up and uses it as an anti-satellite weapon. The weakness of this and similar arguments is that satellites that are in effective orbit need to communicate with ground control. They have what is known as onboard housekeeping systems which every so often have to check in with ground stations. These signals would be intercepted and monitored and so it is very difficult to see how a 'sleeping' ASAT could remain undetected.

However, that is the state of the space art today. It could be that in the near future some way will be found to plant a mine or a sleeper, even close to the high orbiting systems, without the other superpower being certain as to its function.

Of course, much has been made of the idea of using beam weapons (either laser beams or particle beams) to knock out satellites. This again, is one of the concepts that in popular terms would be attractive because it neatly fits the 'Star Wars' atmosphere that many, certainly among the viewing public, would rather hope was a thing of now rather than the distant future. Furthermore, there is also something 'clean' about beam weapons. No explosions, no mushroom clouds and, as with the rest of the elementary concept of 'Star Wars', nobody gets hurt—only the machines die.

Certainly the idea of developing beam weapons is no longer only fantasy. For example, the USA has had for a long time a flying laboratory (plate 9). Long, involved and costly experiments by both the Soviet Union and the United States show that it is possible to manage either laser weapons or particle beam weapons and to eventually use them. Lasers are now in service with both super-powers although not as space weapons. Exactly how far the USSR has advanced in its plans for space-based systems is difficult to tell.

Certainly there was a time when it seemed that the Soviet programme was far ahead of its American counterpart: indeed there were those who argued that the Americans were so far behind that at any time the Soviets could immobilize satellites using a crude beam weapon without the United States being able to respond in kind. The Soviet programme is now thought to be capable of producing some ASAT beam weapon by the 1990s. There has been understandable speculation that the date could be brought forward but it is difficult to see concrete evidence for this. Yet the speculation coupled with US intelligence analysis has served as a pressure if not a catalyst on the varying US programmes. (At times it would appear that they have suffered from lack of constancy in research and development.) Today the United States has two important programmes for laser systems and what is supposed to be an inte-grated programme for particle beam weapons.

The first laser project is working on the idea that it is possible to have visible laser beams based on Earth that could destroy satellites or missiles by either directing the beams on to the target or having an Earth-based beam reflected on to the target by some orbiting mirror. In the longer term the idea would be to put up the whole system into space.

The second project is really a three-pronged attack. Firstly there is Talon Gold, the name given to the research and development programme for the target-tracking system necessary if laser weapons are to work. Lasers and other beam weapons have, in theory, the great advantage of instant sight. Unlike most weapons, they don't have to be aimed at some point, say, ahead of the fast-moving target where the aimer believes or calculates the target will coincide with the arrival of the warhead. In simpler terms, it is rather like a hunter with a shotgun aiming just in front of flying game because he knows that he must allow time for his shot to travel the distance between him and his prey. Lasers do not have this problem. In theory, if they can see the target then they can kill. Talon Gold is trying to take advantage of that speed by allowing the laser weapon to pick up and track the target and then direct its beam onto it. In the second area of research, the Alpha project is trying to develop the long-range lasers that could be used in space. And on the third prong is LODE, The Large Optics Demonstration Experiment. 'Large Optics' suggests one of the problems of laser weapons—the optical technology including mirroring is complicated and in many respects cumbersome. One reason to keep any beam weapon on Earth is that for the moment at least it would need huge power supplies. To get that energy into space would require great effort and even with all the promises of space laboratories and the use of the shuttle to ferry parts into an orbit from which a complex could be constructed, there is some time to go before the technology is at a satisfactory level.

Lasers have another drawback. It is thought that it is possible to provide a more certain defence against them by coating the target in a reflecting or absorbing material; add to this the problem of refraction though uncertain atmospheres and conditions and it is easy to see why scientists are cautious about the claims and hopes of their political masters.

Particle beam weapons have offered scientists a further area for research which started, in the United States at least, during the late 1950s with the SEESAW project (although this was not an ASAT project). SEESAW continued until the early 1970s and concerned itself with the idea that electron particle beams could be used to knock out attacking ballistic missiles. Perhaps the programme was over-ambitious. It had concentrated necessarily on long-range weapons. There was no point in trying to develop a beam weapon that was a last resort against incoming missiles. There were atmospheric barriers to this. A beam weapon is vulnerable to variations in the atmosphere, and in the case of a charged particle beam can bend in the Earth's magnetic field. The technological and financial obstacles were considered too high, and so in 1972 SEESAW was abandoned. It should be noted that the same SEESAW technology for an anti-ballistic missile defence could have been used to build an anti-satellite weapon.

The exotic weapon systems, including those that might one day use neutral particle beams (not liable to deflection by magnetic fields), have a long way to go in research and development. But the feeling among many scientists is that new developments are moving ahead at such speed that it would be wrong to rule out the possibility that these systems could be weapons of the 20th century.

As long as ballistic missile defence is being developed, the corresponding ASAT technology will be available. Perhaps an irony of technology is that those charged with ASAT development take their first look at the problem by examining the most traditional method of attack. Satellites, of course, rely on ground stations. In fact, without ground stations satellites are quite useless. Thus there are those who point out that although hunter-killer satelites and missiles are coming into service, the potential ASAT saboteur has been around for years. It is certainly understood that Soviet Special Forces, known as *Spetznaz*, have reconnoitred many US satellite ground stations. In time of tension the war in space could even be started by a pair of bolt-cutters!

6. Space warfare and beyond

The role of special forces in disrupting the land terminals of space-based communications should not be underestimated. Indeed the British and American governments, among others, have long recognized the urgent need to develop guard forces to protect what are known as KPs (key points) in time of tension. Even in the United Kingdom, where these matters are not discussed openly, there have been meetings at government level to review the ground-based protection systems, a reminder if one were needed of how crucial space systems are to military functions.

Nevertheless, the attention is naturally and realistically diverted to the high ground of space. One of those who supports the concept of space weapons is Paul J. Nahin of the University of New Hampshire. During the 1983 SIPRI Outer Space symposium (published as *Space Weapons — The Arms Control Dilemma*) he put forward the view that 'weapons of mass destruction' should not be deployed in space. Nuclear weapons and other weapons of mass destruction have become unlikely candidates for outer space warfare since their placing there is banned by the 1967 Outer Space Treaty and since nuclear explosions above the ground are also banned by treaty. That's one reason: another is that they are not militarily useful in space. So, Nahin put forward a case for *non-nuclear* systems in space. His argument, adopted by many, was that anti-satellite systems are clearly hostile in intent and therefore could raise tension by denying intelligence during crisis periods. He suggested that non-nuclear technology could be channelled into anti-ballistic missile systems where it might actually contribute to a decline in the arms race (although not everybody agrees with such views). In any case, it is hardly a practical proposal, since anti-ballistic missile weapons can also be used against satellites.

Certainly people such as Nahin have their strong supporters but at this stage in the development of space-based systems the motives of the majority are more complex and perhaps less idealistic than the

71

Nahins of this debate. With presidential backing in the USA and a continuing emphasis in the Soviet Union on anti-satellite systems, it is inevitable that the more aggressive nature of the military use of space is paramount. However, this is no new thing.

The United States Air Force established the Aerospace Corporation during the 1950s. It was put together to supply engineers, scientists and technicians to work on rocket systems. The Air Force wanted the cream and got it. Today the Aerospace Corporation supplies more than 1 000 experts to the US Air Force Space Division. Within the Division there is a pioneer spirit and a sense of elitism. The same atmosphere exists just outside Moscow at the Soviet Union's Space City. There is no guided tour available to Space City, but through contact with some who work there it is easy to spot the similarity between the US and the Soviet space programme people: they are uncluttered by the political and moral issues involved with their work. Tom Karas, at one time a research consultant to the Association of American Scientists, endorses this feeling that the modern spacemen have an almost unstoppable enthusiasm. He points out, for example, that "the spaceman's bible" is a book called *A Few Great Captains*. It isn't a systems folio, nor a directory of procedures; it is a history of "the five Army Air Corps officers who became the patriarchs of the US Air Force".

What is being said in both the Soviet Union and in the United States is that outer space is another military frontier. Scientists, pundits and politicians may debate the moral, financial and practical possibilities, but the people in the industry know that in the long term, no president, no budget and no treaty will prevent space from becoming a military colony. Today nobody questions the idea of having an Air Force. Tomorrow, few will question the existence of a Space Force.

The enthusiasms and predictions, of course, will not hold together a dollar- and rouble-eating space programme for very long. During the autumn of 1983 President Reagan was advised that one aspect of the space programme, future ballistic missile defence (BMD), should be demonstrated in public. Considering that this was the basis of the 'Star Wars' aspect of the President's speech in March of that year, he was inclined to listen. The argument presented to the White House stressed the importance of showing that the USA is determined to explore, and has the competence to develop, the required ballistic missile defence technology. He was told that there could be no reliable overall demonstration within his term of office,

72

even assuming that he ran for and won a second term as President. (At that stage he had not announced his intentions.)

However, the advice was that if sufficient dollars were poured into research and development and with the necessary political backing, then a demonstration of the potential of BMD could be given within 10 years. The report said also that it would be en expensive project. One estimate in Washington suggested that the programme could, at its peak, absorb as much as 2–2.5 per cent of the annual defence budget. Some estimates put the BMD share as high as eight per cent a year for the next five years.

But, what was to be demonstrated? It became clear that a 10-point package was envisaged. The first thing would be to demonstrate that military scientists were thinking along the right lines when they said that it was possible to target a Soviet ICBM during the boost stage and attack it with beam weapons. It would be necessary to adapt the X-ray laser developed at the Lawrence Livermore Laboratory as a space-based weapon. Some reliable system for finding a target, tracking it and then selecting the moment to attack would have to be high on the list of demonstrations.

Some of this could be done within two or three years, but the whole package could not be demonstrated collectively until the "early 1990s". The recommendations did not silence the critics of the beam weapon programme. The fantastic opportunities suggested were a little too hard to swallow for many of a more sombre scientific temperament. Paul Nahin had already dismissed those doubts during the 1983 SIPRI Outer Space symposium, which took place *before* the report. Nahin argued that he believed that it would be possible to destroy a single, unprotected ICBM after its launch and before it had time to release its warheads; that it was (or eventually would be) possible to do so with beam weapons; and that there may still be those who will deny this, but they always end up supporting their arguments by reciting limits of present-day engineering. Such limits, he claimed, are nothing but limits of the imagination. As a reminder he recited Clarke's Law:

> If an element authority claims something is possible, he may be wrong but most likely he is right. If, however, he claims something is absolutely impossible, he may be correct but most likely he is wrong.

But given the safeguard of not falling foul of Clarke's Law (named after Arthur C. Clarke) and full technical, financial and political endorsement, would it really be possible to defeat or even deter a major ballistic missile attack from space? Could space really become a credible battleground?

There have been proposals for space stations in permanent orbit fitted out with laser weapons. Other plans involve space mirrors able to reflect beams from land-based weapons on to targets. Counter-arguments suggest it would be simple to launch dummy missiles which would confuse and exhaust the space stations. They would not know which, if any, target was hostile, but would have to take on each one. Eventually, when the space battle station was exhausted, then the real ICBMs would be sent in.

As Nahin admits, an orbital system could be defeated by inserting clouds of junk into the same orbits as the battle stations. The kinetic energies of impact would shred the battle stations and it would mean that such stations would have to be given sophisticated control features that would allow them to out-manoeuvre the clouds of junk. It should be remembered that an effective BMD system would be very complex and very costly. Countermeasures to neutralize such systems may be relatively easy (and some would say relatively cheap).

The 1983 report to the President did suggest that current technology could counter existing Soviet missiles. But there seemed to be a belief that advances in Soviet technology would soon erode this advantage. In this simple assessment there is an acknowledge-ment of an arms race that will be very difficult to stop. Indeed, the very demonstration called for by the report would necessarily en-courage the Soviet Union to develop more advanced and presumably less vulnerable missile systems.

One lobby for US anti-ballistic missile beam weapons claims that the opposite is true. It suggests that if the United States produces and shows it has the capability of stopping ballistic missiles, the Soviet Union will see it as a hopeless position and develop fewer offensive weapons and concentrate its resources on defensive systems. The past history of the arms race, however, provides no support for this view.

The US journal *Aviation Week & Space Technology* concluded that the programme being recommended to the President is designed to increase the Soviet role in co-operating with a stable strategic environment. It is also designed to "alter the Star Wars fantasy, establish the credibility and then the reality of defensive technologies...". The journal article proposes rapid development of technology that can deal with postulated Soviet countermeasures.

74

This development approach would also provide a hedge against similar Soviet defensive technologies.

At this stage it would be prudent to go along with the philosophy that insists that although we don't know if this aspect of the military use of space would work, it is better to assume that at some time in the not too distant future some form of the technology will be attempted and *might* work. As Nahin reminded the 1983 SIPRI symposium:

> Vannevar Bush, one of America's most honoured and inventive men of science, made the famous error of stating (in 1945) that the inter-continental ballistic missile was impossible— Bush was in good company, of course, as it was Ernest Rutherford who, in 1933, warned nuclear energy proponents that their hopes were "the merest moonshine".

So, what do today's proponents of BMD hope to achieve in their designs on outer space?

7. Stopping ballistic missile attack?

Let us not confuse the ambitions of, say, the sentiments expressed by the White House and the more realistic sights set by the scientists. The United States has, as mentioned earlier, seen for some time the possibilities of being able to stop or at least blunt a ballistic missile attack by using space-based systems. The 'first stage' was officially launched by President Reagan's 'Star Wars' speech of March 23, 1983. The 'booster' was fired by his State of the Union speech of January 27, 1984.

During this latter address, President Reagan said that he had directed NASA to take what was described as the nation's "next bold step" in space: to begin the development of a permanently based, manned station. It was announced later: "The President's goal for the space station program is to have Americans living and working in space, permanently, within a decade".

There was no indication that the space station programme related to the ballistic missile defence ideas voiced 10 months earlier. However, there were those in Washington who acknowledged that a permanently manned "space capability" was "desirable" if the full potential of any BMD research was to be fulfilled. One aspect of space-based systems is, of course, high-energy beam weapons. The US Department of Defense plans to invest some $24 billion on strategic defence between 1985 and 1989. Of this some $5.5 billion has been requested for the development of beam weapons.

Of course, the military is not pinning all its hopes for ballistic missile defence on beam weapons or even on other space-based systems. The idea is to have a three-layered defence, with beam weapons as candidates for the first layer (see figure 18).

The technologies being considered for the other two layers of defence are much less in the realms of science fiction. Many are already being tested and promise, if not to actually stop a missile attack, to blunt its ferocity somewhat. The funding for research and development for these two layers has been stepped up dramatically.

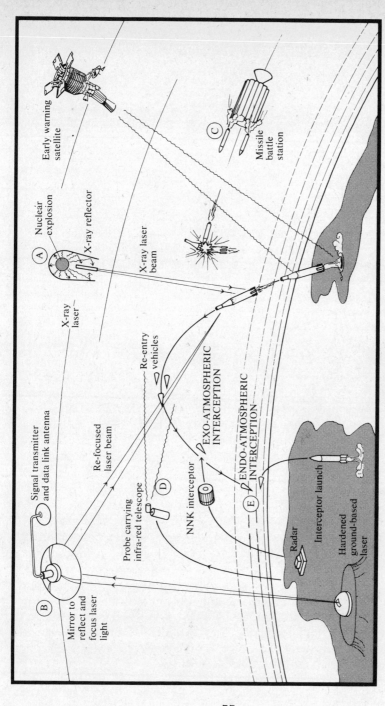

Figure 18. Is this how the US BMD system will look in the year 2000? Essentially, three layers of defence are planned. A missile is detected by an early warning satellite and intercepted during the boost-phase either by a nuclear explosive X-ray laser (A), by a ground-based laser beam reflected off a space-based battle station (C). The missiles which escape destruction release their warheads which are detected by an infra-red telescope and intercepted by non-nuclear (NNK) warheads above the atmosphere (D). Those warheads which escape this interception are detected by ground-based radars and intercepted by conventional nuclear warheads (E), this time within the atmosphere.

For fiscal years 1982, 1983 and 1984 funding was just over $460 million, $519 million and $660 million respectively. Then for fiscal year 1985 $1 293 million was requested, indicating that there is also an active interest in less exotic forms of BMD.

How would a three-layered BMD system work? Lasers, either space-based or land-based, are the most popular candidates for the first layer, one reason being that a laser beam is not bent by the Earth's magnetic field. Once a warning of attack has been received, probably from an early warning satellite, mirrors would be launched, according to one concept. These mirrors, once in position, would focus laser beams emitted from ground-based lasers onto the ballistic missiles after they had emerged from the atmosphere. Laser battle stations already in space, according to another concept, would unleash their destruction instantly on warning. Thus ICBMs could be killed within about three minutes of being launched. Lasers based in space have another advantage over ground-based lasers—minimal problems of atmospherics.

The military attraction to attacking ICBMs in their boost phase is threefold. At this early stage the missile still has all its multiple warheads so it's a matter of killing up to 10 'birds' with one 'stone'. Secondly, because of the hot exhaust gases the missile is easily detected and tracked. Thirdly, in the boost phase the missile still has all its fuel and is under enormous strain, so it can easily be exploded.

Another idea for the first layer of the defence is the missile-armed battle stations proposed by the High Frontier group. The idea is to have 432 satellites, each armed with 40–45 missiles, permanently in orbit. Each of these heat-seeking missiles would home in on enemy ICBMs and destroy them by collision. This idea has attracted considerable interest.

Missiles that escape the first layer of defence will eject their multiple warheads at the height of their trajectory. These would now face the second layer of the defence operating above the atmosphere. This second layer is based on new though conventional technology and is well on the way towards perfection. Interestingly, the spearhead of the second layer would be a ground-launched non-nuclear warhead similar to the miniature homing vehicle of the new US ASAT system (which further underlines the similarity of BMD and ASAT technologies); in fact, the BMD warhead was developed first. It would operate in conjunction with an infra-red telescope device that would detect and track the incoming nuclear warheads against the cold background of outer space.

The third layer of the defence would operate within the atmosphere against nuclear warheads that had leaked through the other two layers. This technology exists but is being improved and non-nuclear systems are being investigated. Nuclear-armed ABM missiles were deployed on a limited scale in the early 1970s as allowed for under the 1972 Anti-Ballistic Missile (ABM) Treaty: they protected missile silos at Grand Forks, North Dakota. But they were decomissioned in 1975 because the nuclear warhead brought numerous complications including the fact that use would destroy US communication systems.

That is basically the grand design for US BMD. An equal amount of information is not known about Soviet plans. However, there are traditional BMD systems deployed to protect Moscow as allowed for by the ABM treaty. The Soviets certainly have an active programme for improving these ballistic missile defence systems. This may even include high-energy beam weapons.

We have discussed these 'Star Wars' weapons throughout the book but have not yet said what a laser is. To put it very simply, a laser is usually a beam of light consisting of light rays of the same frequency, in phase, and travelling in the same direction. To produce laser beams, energy is given to atoms and molecules in what is known as the lasing medium. This energy stimulates other excited atoms and molecules in the lasing medium which then emit light in the same frequency, in the same direction and in the same phase. This 'monochromatic, coherent' light is then amplified using mirrors (not the same mirrors that would be launched into space to focus land-emitted lasers). (A particle beam, which is intended more for offence, is a flow of sub-atomic particles energized in a device known as an accelerator. The particles can be electrically charged or neutral.)

High-energy lasers are those which have an average power output of more than 20 kilowatts (kW). Such lasers are being used in industry. For beam weapons applications lasers are thought to need energies ranging from a few hundred kilowatts to many megawatts. The US airborne laser laboratory is equipped with a 400 kW laser (see plate 9).

In a laboratory environment, a 10 kW laser can easily penetrate a few centimetres of steel in a fraction of a minute while, theoretically, a 5 MW hydrogen fluoride laser can make a hole through steel 0.2 cm thick in less than 10 seconds from a distance of 1 000 km. In the latter case it is assumed that the target surface is non-reflective and has not been protected in any way and that the laser beam has been focused to a minimum radius of about one metre.

In February 1981, the US Air Force tested its aircraft-mounted laser system at full power on the ground. Later, on June 1, the airborne laser was tested against an air-launched AIM-9L Sidewinder air-to-air missile. While the beam hit the target, it did not destroy it. Two days later a second test was carried out against an AIM-9L missile. It was able to lock onto the target for a long period. More recently, it was reported that in July 1983 the laser beam engaged five Sidewinder missiles and managed to change their course.

The US Navy, under a programme called Sea Lite, is investigating a more powerful chemical laser. The Defense Advance Research Project Agency is investigating in ground tests the feasibility of deploying such a laser in outer space.

Several types of high-energy lasers have been proposed for weapon applications. For various reasons, short-wavelength lasers are preferable. Nonetheless, considerable effort has been devoted to research on long-wavelength laser devices. For example, the US Air Force airborne laser laboratory is equipped with a gas dynamic laser (GDL) which uses carbon dioxide. In a GDL, the rapid expansion of a gas provides the initial energy input, while in a chemical laser the corresponding conditions are achieved by means of chemical reactions. The most commonly used chemical reactions for the latter type of laser are between hydrogen and fluorine or deuterium and fluorine.

In the electric discharge laser (EDL) the lasing material is excited by collisions with the electrons of an electric discharge. Another type of EDL is called the excimer laser which operates in the visible and ultraviolet wavelengths. This latter laser type has produced high-power bursts at short wavelengths in the laboratory.

A relatively new and, in principle, different and tunable laser type is the free-electron laser (FEL). It might be developed to exhibit a very high efficiency in converting electrical energy to laser energy.

Other potential devices which have entered the high-energy laser weapon debate are the gamma-ray laser or 'graser', and the X-ray laser. These have been subject to theoretical analyses both in the USA and USSR for more than a decade.

An X-ray laser can in principle be pumped by a high-intensity flash of X-rays from a conventional X-ray source. However, these are generally not intense enough. Since X-rays can't be amplified using mirrors this necessitates the use of copious radiation from a nuclear explosion as the pumping source of the X-ray laser. It was reported in 1981 that the Lawrence Livermore Laboratory had tested the concept of the X-ray pumped laser during an underground nuclear explosion. A second test of this, the so-called 'Excalibur' (X-ray laser), was also reported and more are planned.

While the US Department of Defense has so far concentrated on the chemical laser for a possible space-based system, the emphasis seems to be shifting towards free-electron, excimer, graser and X-ray lasers owing to the limitations of chemical lasers. One reason is the very large amount of fuel needed for chemical lasers, which must be transported to the orbital laser platform.

Among important questions of a technological nature which remain to be solved is the need for a compact power source to supply input energy for a laser. The Soviet Union has been orbiting small 10 kW(e) nuclear power reactors for the past decade or so and the USA has a plan to orbit a 100 kW(e) reactor in the very near future. In any case, if grasers and X-ray lasers are to be made to work, they may derive their energy from small nuclear explosions. The other problem is that of finding, aiming at and tracking a target. Some of these problems are common to other areas as well; for example the NASA space telescope and space surveillance and anti-satellite activities. Once these problems are solved the techniques will, no doubt, be applied to laser weapons as well.

Thus there are several technological problems to be solved before some of the BMD systems can be realized. However, a decision to develop a space-based laser BMD system has considerably far reaching implications. These will be discussed in the next chapter.

8. What are the implications?

In space exploration there has been a need, in the minds of the military, to take advantage of this new arena of scientific learning. This need is not a sinister trend. That would imply that the military has no right to be in space. To argue that the military community should ignore what we have chosen to call the new high ground would be intellectually naive. Yet it is not enough to accept the next stage of the argument: that because the military influence on space exploration is so great, nothing should be attempted to control that influence.

Perhaps then, the role of the academic, the politician, the diplomat and the international legislator is to put forward sound thinking that would realistically preserve outer space for peaceful purposes so that it does not become some trigger for major conflict. The preceding chapters have identified the dangerous elements and trends; therefore it is worthwhile, in this final chapter, to go over those elements and put them in today's context, before proceeding with suggestions that might be considered if the world is to become a safer place in which to live.

Those who consider themselves to be well versed in the history and development of space exploration and the dangers that have followed will surely be tolerant if they feel that much of what follows is an essay or a précis of what has gone before. It is the experience of the authors that even the simplest of long explanations needs a summary; furthermore, we hope that this last chapter might be a ready reference for those who wish to argue the case for arms control in space.

THE BEGINNING

As both superpowers circled the Earth, there must have been an instinct in mere Earthlings that wondered, even hoped, that this was the start of an era that would bring first the Moon and then the planets within reach. Would it be possible to land on Mars? Was there extra-terrestrial life waiting to make contact? It was, naturally, an exploration of hope, not of doubt.

And yet to many, like the authors, who watched from this Earthly mothership, there was a sense of foreboding eroding the senses of wonder and admiration during that period between the late 1950s and early 1960s. They recognized that man didn't have to go to the planets to wage space war or to threaten the peace of the Earth. It was sufficient to go into near-Earth orbits to establish comprehensive reconnaissance systems and, in terms of space technology, not that much higher to provide the basis for an extensive communications network and even one day, a space-based weapon system. Today, that capability, and in some cases that actuality, is with us.

THE SYSTEMS

From 1958 to 1983, 2 114 satellites had been launched that had some military use. In common terms that would mean that every three days sees the launch of some military spacecraft. Consequently, 75 per cent of all satellites have some military value.

Earth-orbiting satellites are used by the military to enhance the performance of Earth-based armed forces and weapons. Such military missions range from navigation, communications, meteorology and geodesy to reconnaissance. The main function of reconnaissance satellites (photographic, electronic and ocean-surveillance spacecraft) has been to obtain information on military targets.

Photographic reconnaissance satellites detect, identify and pin-point military targets. In addition to photographic cameras, sensors include television cameras, multispectral scanners and microwave radars. Some of these instruments can spot objects as small as 30 cm in diameter. Both the Soviet Union and the United States launch such satellites regularly, and the People's Republic of China has launched at least five. Of the military satellites launched by these nations, some 40 per cent have been used for photographic reconnaissance purposes. France has plans to launch such a military satellite, called SAMRO (Satellite Militaire de Reconnaissance Optique), in the late 1980s. Japan may also be interested in developing a military reconnaissance satellite.

Electronic reconnaissance satellites are the 'ears' in space. They carry equipment designed to detect and monitor radio signals generated by the opponent's military activities. Signals originate, for example, from military communications between bases, from early warning radars, air defence and missile defence radars or from missiles during tests. These satellites also gather data on missile testing, new radars and many other types of communications traffic. Not only do they locate systems producing electronic signals but they also measure the characteristics of the signals so as to be able to plan penetration of defences.

Ocean-surveillance and oceanographic satellites detect and track naval ships and determine sea conditions which can, for instance, help in forecasting the weather or, less innocently, in detecting submarines. Space-based sensors include radars that can 'see' through clouds and detect even small pleasure boats. The radars on Soviet spacecraft are powered by small nuclear reactors.

Knowledge of what is happening in the oceans—for example, the height of waves, the strength and direction of ocean currents, and salinity of the sea water—can help in the design of sensors to determine whether submarines are lurking beneath the surface. These factors also contribute to improving the accuracies of missiles launched from submarines.

Early warning satellites have partially replaced the radars that were originally deployed to give warning of a surprise attack by ballistic missiles. The radars provided about 15 minutes for a response to be worked out. The use of early warning satellites has extended this warning time to some 30 minutes. The US early warning satellites are placed in geostationary orbit some 36 000 km from Earth while such Soviet satellites move in a highly elliptical orbit. The more recent versions of the US Rhyolite early warning satellites (versions were sometimes renamed Argus, and some reports suggest that the name was changed to Chalet) are also 'ears' in space; they perform mainly electronic reconnaissance, particularly monitoring telemetric signals emitted from missiles during their test flights. Some of these satellites carry additional sensors to detect nuclear explosions conducted above ground, in the atmosphere and in outer space.

Communications satellites are beginning to fulfil the military need for rapid and efficient communications to aid highly complex and sophisticated weapons. Moreover, space-based sensors for surveillance of Earth, together with land-based surveillance systems, generate a considerable amount of data. The transmission of this and other data for military purposes needs reliable and secure communications systems. Space has become an area of vital interest as, for example, 80 per cent of US military communications are carried out using artificial Earth satellites. These satellites also play a vital role in the command and control functions for the military forces of the great powers. Even communications between mobile forces such as aircraft, naval ships and soldiers on foot and their commanders is being conducted via satellites. Moreover, the communications between the two superpowers in a state of emergency would be via satellites.

Navigation satellites are a family of military craft which send out coded signals with which armed forces can plot their own positions with a high degree of accuracy. For example, the planned 18-satellite navigation system of the USA, the Global Positioning System (GPS) or NAVSTAR, will determine the position, in three dimensions and to within about 20 m, of an aircraft, missile, naval craft, or soldier on the ground.

Both the USA and the USSR have developed satellite navigation systems. In the USA, an added function is planned for such satellites. Although the USA has launched satellites specifically to detect nuclear explosions in the atmosphere and in outer space, it is now planned that US navigation satellites will carry sensors for this purpose. This is intended also to provide damage assessment both within the USA and within enemy territories during and after a nuclear attack. This effort is in support of the nuclear war doctrines which require early warning of attack, information for assessing the size of the attack, and data on the attacked target so that an appropriate response can be made.

Meteorological and geodetic satellites form the final members of the military spacecraft armoury. The former can gather information about the weather along a missile's proposed route so it can be guided accurately. The latter obtain data about, for example, the shape of the Earth or its gravitational field to achieve the same result.

The amount of data collected by meteorological satellites is considerably more than just to know whether an area of interest is covered by clouds in order to plan photographic reconnaissance or bombing missions. For example, there are sensors on board which measure oxygen and nitrogen density, and which determine the temperature and water vapour at various altitudes. One reason for such detailed measurements of the properties of the atmosphere could be that once man has understood the mechanics of weather and climate formation, his military genius may be able to control these for hostile purposes.

However, an immediate application of such data is for improving the accuracy of missiles. Among the factors influencing accuracy are the water vapour content in the atmosphere and the wind velocity along the missile's possible trajectory. Not only do the weather conditions determine the corrections to be made to missile trajectories but these conditions are also taken into account when predicting satellite orbital tracks.

SPACECRAFT AND THE RISK OF WAR

A review of the military encroachment on outer space indicates the extent to which the military satellites of the two superpowers are becoming part of the world-wide nuclear and conventional weapon systems. As advances are made in military space technology, improvements in the efficiency of weapons occur. This in turn may refine war-fighting tactics as well as give rise to new ones.

Over the past two decades or so nuclear doctrines have moved away from mutual assured destruction (MAD). According to MAD, massive destructive power would be unleashed against enemy cities and industrial centres in response to an attack; it was assumed no side would attack first for fear its own cities would be destroyed. This was a policy of deterrence.

But new doctrines have come into being that actually consider fighting a nuclear war. The assumption is that it can be controlled—limited in intensity and area—hence all the talk of a limited nuclear war in Europe. The military capability to back up these ideas could hardly have been acquired without the help of space technology.

These nuclear war-fighting strategies demand, for example, pin-point accuracy to limit damage to military targets (remember the role of navigation, geodetic and weather satellites in improving missile accuracy). Among other needs is a secure C^3I system. Increased discrimination in targeting could be obtained from accurate knowledge of targets as well as through all target information being available—instantly—at headquarters. A side must be able to know if the target is destroyed and if not strike again—immediately. In all these tasks communications and reconnaissance satellites play a vital role. These are just a few examples of ways in which space has helped to make possible the idea of limiting a nuclear war.

But is that so bad? One might ask if it isn't preferable to threatening the lives of millions of civilians. One problem is that the increased nuclear war-fighting ability could lead a side to believe it can actually win a nuclear war by destroying in a first strike the enemy's ability to counteract. Or it could lead to a situation where a side strikes first to prevent its own arsenal being destroyed in one massive strike by the other side. Another danger is that the belief that one can limit a nuclear war might make nuclear war more probable.

Be that as it may, in view of the potential role of satellites in waging wars on Earth, it is not surprising that they have become important military targets. Thus began what could be called the second phase in the militarization of outer space. That is, the development and even in some cases the deployment of anti-satellite (ASAT) weapons, a phase which unless checked could lead to war in space—and which might trigger nuclear world war.

The US ASAT weapon based on the Thor missile was first tested in 1964. This and a similar system using the Nike-Zeus missile deployed nuclear warheads. These systems were dismantled in 1975 but interest in ASAT weaponry was revived in the late 1970s, which led to the present multi-faceted ASAT programme. The new American ASAT weapon is known as the miniature homing vehicle (MHV). This will be fitted to a two-stage rocket launched from an F-15 aircraft (see plate 10). The USA carried out its first test of the missile launched from the F-15 (without the MHV) on January 21, 1984. The system may have a range of up to 460 km.

The Soviet conventional ASAT programme may have begun as early as 1963. Basically, a satellite is launched to hunt down a target satellite as it orbits the Earth. The hunter-killer satellite is either exploded using conventional explosives or after coming very close to its prey is commanded back to Earth.

It should be noted here that a hunter-killer attack is slow. The US air-launched ASAT missile may be much more efficient, but targets in geostationary orbit will still be difficult to reach. This is why the more exotic systems, such as ground- or space-based high-energy laser beams, have recently attracted the attention of the military as possible ASAT weapons.

Obviously these ideas are related to the newer concepts of **ballistic missile defence** (BMD). Sometimes they are called Star Wars systems, although the US Department of Defense has said that it does not like this title and would prefer to talk about 'strategic defence'. Whatever the terminology, it is clear that the summary contained in this chapter and the description of ballistic missile defence in the previous chapter indicates what we believe to be the military use of outer space. Realistically there is very little that any agency can do to remove this capability. No one country is going to immediately pull down their considerable space-based military systems for any treaty, no matter how perfect the wording. International co-operation and diplomacy simply does not work that way. However, that does not mean that some control should not be established and restrictions put forward that are workable and reassuring.

ARMS CONTROL

The treaties that we now have are not adequate for a technology that has moved forward so rapidly that it has outstripped the intention and spirit of existing treaties.

The 1967 Outer Space Treaty deals very inadequately with the problem of the militarization of outer space. The only limitation placed on military activities in this environment is the prohibition of the placing in orbit of nuclear weapons or other weapons of mass destruction, which are very doubtful from the point of view of their military effectiveness. The treaty thus legitimizes other military use of outer space. The Moon and other celestial bodies are demilitarized, but outer space as a whole has only been partially demilitarized. With the exception of the 1963 Partial Test Ban Treaty, other subsequent treaties and new proposals have legitimized the military use of outer space and, in fact, offer protection to military satellites. For example, the 1972 Anti-Ballistic Missile (ABM) Treaty and the SALT I and II treaties contain pledges not to interfere with the ''national technical means of verification''. This means that effectively photographic, electronic and early warning satellites as well as some ocean-surveillance satellites are protected by these treaties from destruction or interference. The 1973 International Telecommunications Union Convention obligates parties to the Convention to avoid harmful interference with the radio service or communications of other parties.

Thus, only military navigation, geodetic and meteorological satellites are without any protection. But when devices to detect nuclear explosions are deployed on board the US NAVSTAR satellites, these might also be classified under the category of national technical means of verification and so be protected by some of the above-mentioned treaties.

A number of arms control measures in outer space have been proposed both officially and by non-governmental organizations. These are further attempts to protect military satellites, to a varying degree.

A draft treaty proposed by the Soviet Union in August 1983 apparently bans all ASAT weapons, but the proposal is not without ambiguity. For example, article 1 prohibits the use or threat of force through the utilization of space objects in orbit. It further prohibits the use or threat of force *against* space objects. This may be interpreted as allowing possession of ASAT weapons.

The authors' firm belief is that possession of ASAT weapons should be banned because in time of crisis a side with weapons will be tempted to use them. Also, there is a very real chance of a satellite being accidentally damaged. In a time of tension such an accident might be interpreted as an attack if it was known that the other side had ASAT weapons. This seems all the more likely when one considers that some vital early warning satellites are found in geo-stationary orbit, which is already so crowded that it approaches saturation point. Collisions have only narrowly been avoided.

One interesting feature of the Soviet proposal, however, is that it includes a prohibition on space-based ballistic missile defence, emphasizing the dual nature of BMD and ASAT systems.

Even though the technological problems relating to a space-based BMD laser weapon system may not be solved in the foreseeable future, this proposed application of high-energy laser beams raises considerable difficulties from the point of view of arms control. For example, the possible use of high-energy beam weapons as a BMD system may have a destabilizing effect on the relationship between the two superpowers. If one side had such a system, it might be tempted to strike first against the other, probably using short-range nuclear weapons, believing it could still defend itself against its opponent's ballistic missiles. The result would be a gradual involvement of nuclear weapons of all kinds—escalation.

Another consequence would be for both the USA and the USSR to embark on yet another round of the arms race. Not only would there be the obvious—and costly—competition in laser BMD but one of the reactions might be for both sides to multiply drastically their offensive nuclear arsenals. This would be to ensure that despite the opposing BMD some nuclear weapons would reach their targets. Thus, far from rendering nuclear weapons obsolete, lasers might actually accelerate the nuclear arms race.

Another serious implication of these weapons is that their development violates the spirit of the Anti-Ballistic Missile Treaty. The idea at the time (1972) was that the ABM Treaty represented a commitment by both sides to remain vulnerable in order for deterrence to work along the lines of MAD. But the new war-fighting strategies demand that weapons should survive attacks. Hence one can see that the motivation for BMD development runs deep. The continuing advances in defensive weapons may well create a pressure to abrogate the ABM treaty.

There is an arms control difficulty raised by the new technology which has been discussed very little—if X-ray lasers are deployed this may jeopardize the 1963 Partial Test Ban Treaty which bans nuclear weapon tests in outer space, in the atmosphere and under water; as we have said, X-ray lasers can be produced using nuclear explosions and would, of course, not be deployed without considerable testing.

Certainly the deployment of X-ray lasers would violate the 1967 Outer Space Treaty, which prohibits orbiting nuclear weapons and other weapons of mass destruction. In any case, the Outer Space Treaty would be violated in spirit since the Treaty requires space to be used for peaceful purposes only. Orbiting a BMD system cannot be regarded as a peaceful activity since it can also be used for offensive purposes, that is, as an ASAT weapon. This dual use brings another problem.

Plans for a ballistic missile defence system that can also be used as an ASAT weapon will complicate discussions on a possible anti-satellite treaty.

Thus the implications of the continued militarization of outer space are complex and far-reaching. Any consideration of arms control measures will have to be equally far-reaching—we discuss such measures in the Recommendations (page 93).

Now might be the time to emphasize the more positive aspects of space technology and to broaden international co-operation in this field. The single most desirable military use of space is verifying that the superpowers are keeping to the terms of some arms control treaties. With this in mind France, in 1978, proposed the establishment of an international satellite monitoring agency (ISMA). The agency would use satellites to verify multilateral arms treaties as well as to monitor crisis areas. The idea was that space should be used to enhance confidence and security among nations.

Today only the superpowers are able to gather all the sensitive data relevant to verifying arms control agreements. Confidence would be increased by an ISMA because other members of the international community would be able to use satellite information to check that *multilateral* treaties were not being violated.

Some of the main impetus for an international satellite monitoring agency lies in the fact that an increasing number of countries are developing their own launchers and satellites. While this trend is mainly limited to industrialized nations, a number of other nations are acquiring data-receiving and image-processing technology. Moreover, the performance of civilian satellites is improving rapidly (see, for example, plate 7).

There are two major obstacles to an ISMA. Firstly, the super-powers are unwilling to give up their long cherished monopoly over the technology. Secondly, there are sensitive security considerations of states. Also satellite pictures would reveal oil, mineral and agricultural data which could be used to wage economic wars.

While discussions of such problems must continue and efforts must be made to solve them, it may be useful to think of satellite monitoring on a *regional* basis.

As we have pointed out both superpowers have monitored many conflict areas from space. They may also be observing military manoeuvres in Europe as a confidence-building measure. This may be very relevant to current negotiations such as the Stockholm Conference on Confidence- and Security-Building Measures in Europe. Discussions at Stockholm include measures to reduce tensions caused by manoeuvres. When foreign observers are present at manoeuvres their observations can only be limited, particularly in the case of a large manoeuvre spread out over a great area. Thus Europe could be identified as the first region for a regional satellite monitoring agency (RSMA).

This is particularly interesting since the *infrastructure* needed for an RSMA already exists in Europe. For example, in Western Europe there is the European Space Agency (ESA). And the Interkosmos Council has been established in Eastern Europe. Both these organizations have very active space programmes, especially in the field of remote sensing—an essential technology for verifying arms control treaties. And very important links between the two organizations also exist. For example, France has actively partici-pated in the Interkosmos programmes of orbiting scientific experi-ments on board many Soviet Cosmos satellites. Also, French astronauts have flown on Salyut Soyuz spacecraft and France is the most active member of ESA.

The problems raised in the case of an ISMA by the availability and spread of sensitive data may be less intense in the case of an RSMA.

For concepts of ISMA and RSMA to work there must be some co-operation between nations. International co-operation concern-ing activities in outer space is not a new concept. Many West European states are co-operating under the umbrella of the European Space Agency. The Interkosmos Council was established in order to further co-operation in space technology among East European nations. Beside these, there are a number of successful international ventures in the field of communications, navigation and in meteorology. The World Meteorological Organization has introduced a global programme, the World Weather Watch, which is actively supported by the European Meteorological Satellite Programme and by programmes of Japan, the USA and the USSR.

Then there is a more recent co-operative venture, an international search and rescue project using satellites, involving Canada, France, Norway, the UK, the USA and the USSR. China, Finland and Japan are negotiating to join the programme. As part of this project, three satellites, Soviet Cosmos 1383, Cosmos 1447 and US NOAA-8, have been launched; several lives have already been saved through, for example, pin-pointing the position of crashed aircraft using Cosmos 1383.

This and other international ventures have been highly successful. It must be remembered that even rivals in arms are co-operating when natural disasters occur; surely they will in the end co-operate in averting man-made disasters?

Recommendations

It is obvious that there is a dilemma. Military spacecraft have contributed considerably to the formulation of strategies for actually fighting nuclear war rather than just deterring it. On the face of it it would seem a reasonable idea to ban all military satellites. But would it? These same satellites can be used to monitor treaties safeguarding international security; they can be used to maintain military communications so that war by miscalculation is less likely—but some form of treaty is needed to protect satellites for peaceful and stable use. There must also be other ways of reducing the destabilizing effects of the military use of space.

Considering this the following points are offered for inclusion and consideration in any debate that may lead towards a draft proposal for the protection of outer space vehicles.

1. While all the satellites launched are registered with the United Nations in conformity with General Assembly Resolution 1721 B (XVI), the information given on most spacecraft is scanty. Important details, such as the launch time and site, physical characteristics of satellites, degree of manoeuvrability, satellite lifetime and often the orbital characteristics of satellites, should be given. At present, the purpose of a spacecraft is often difficult to determine.

2. It is important to closely examine the special body of law dealing with activities in outer space so that there is a clear understanding of which satellites really are protected and what further measures are needed.

3. The minimum that the two major powers could do to bring at least some military activities in outer space under control is to agree to an ASAT treaty that will ban not only tests but also possession.

Unless an ASAT treaty is achieved soon the development and deployment of some of the space weapons which could be used for both offensive and defensive purposes could mean an abrogation of the 1972 Anti-Ballistic Missile Treaty. It would probably also involve breach of the 1967 Outer Space Treaty (which is already being violated in spirit) and even the 1963 Partial Test Ban Treaty. New legal instruments submitted as drafts by the USSR and some non-governmental organizations should be taken as a basis for discussion even on an international basis.

A draft multilateral treaty to control space weapons was proposed by the Soviet Union in 1983. While the US Administration has indicated that it will not seek a comprehensive ASAT treaty, mainly because of the verification problems, it may be prepared to consider a limited ASAT treaty. For example, the Administration may consider prohibiting some specific types of weapons or actions.

4. As an interim measure countries could impose a moratorium on the testing of ASAT weapons and could also make a declaration of no-first-use of ASAT weapons. In fact, the USSR and the People's Republic of China have made a declaration not to be the first to use nuclear weapons. There is a connection in that certain types of laser weapons are regarded as third generation nuclear weapons. A declaration of no-first-use of ASAT weapons may be useful in the case of an incomplete ASAT treaty.

5. The same technologies can be used in both anti-satellite systems and ballistic missile defence. Thus it is no use banning one of these systems and letting the other go ahead. If there were only a ban on ASAT systems, the result would probably be that the technologies which were being developed for ASAT purposes would acquire a new label: we would be told that they were being developed for anti-ballistic missile purposes. It follows that any action against the development or deployment of anti-satellite systems should be accompanied by equivalent action against anti-ballistic missile systems. This may require some renegotiation of the Anti-Ballistic Missile Treaty to ensure that it is water-tight.

6. Since disarmament is a concern for all, it is essential that discussions on the concept of an international satellite monitoring agency are kept alive. A regional satellite monitoring agency can be considered as an interim measure.

7. To ensure availability of satellites during a time of crisis, both the USSR and the USA intend to station spare satellites in orbit. Moreover, for a continuous world-wide availability of accurate navigation capability, it is necessary to have a number of spacecraft in a space-based navigation system. Similarly, a world-wide communications system needs at least three satellites in orbit. It may be suggested that the number of military satellites launched per year by a country should be limited, so that the performance of some

weapon systems on Earth could at least be degraded and even the efficiency of performance or effectiveness of military forces could be reduced.

It is crucial to obtain a clear understanding of the military uses of outer space with a view to clearly defining which of the military satellites are to be permitted and for what purposes. While there are still only two countries which are able to enhance the performance of their nuclear arsenals using space technology, a truly comprehensive outer space treaty containing real disarmament measures must be worked out. Another opportunity will be missed if greater control over the militarization of space is not brought about now.

Further Reading

Secret Sentries in Space
By Philip J. Klass (Random House, New York, 1971).

Anti-Satellite Weapons: Arms Control or Arms Race?
(Union of Concerned Scientists, Cambridge, MA, 1983).

High Frontier
By Daniel O. Graham (The Heritage Foundation, Washington, D.C., 1982).

Outer Space—A New Dimension of the Arms Race
Edited by Bhupendra Jasani (Taylor & Francis, London, 1982)
[a SIPRI book].

Outer Space—Battlefield of the Future?
By Bhupendra Jasani (Taylor & Francis, London, 1978) [a SIPRI book].

Red Star in Orbit
By James E. Obert (Harrap, London, 1981).

Space
By Martin Ince (Sphere Books, London, 1981).

Space War
By David Ritchie (Atheneum, New York, 1982).

Space Weapons: The Arms Control Dilemma
Edited by Bhupendra Jasani (Taylor & Francis, London, 1984)
[a SIPRI book].

The Final Decade
By Christopher Lee (Hamish Hamilton, London, 1981; Sphere Books, 1983).

The New High Ground
By Thomas Karas (Simon & Schuster, New York, 1983).

Index

99

Computers, 51
Conrad, Astronaut Charles, 20
Corporate, Operation *see*
    Falklands/Malvinas war
Cosmonauts, 14, 22, 26–7 *see also*
    under names of
Cosmos series, start of, 14
Cunningham, Astronaut Walter, 20
Cyprus, 25, 26

Deterrence, 86
Digital data, 51
Disarmament and arms control
    treaties, space and, 88–92,
    93–5:
    satellite verification of, 4–5, 53,
    88, 90–1, 93
    see also ABM Treaty, Outer Space
    Treaty, Partial Test Ban Treaty
Discoverer satellites, 8
Discoverer 13, 50
Dobrovolsky, Cosmonaut Georgy, 23
DSCS (Defense Satellite
    Communications System), 46–7
Dyna-soar project, 14

Eagle, 20
Eisele, Astronaut Donn, 20
EORSATs, 40
Europe, nuclear war in, 86
European Meteorological Satellite
    Programme, 92
European Space Agency (ESA), 11,
    91, 92
Explorer 1, 7

F-15 aircraft, 65–6
Falklands/Malvinas war, 40, 41, 42,
    54–5
False colour photography, 51
Faslane, 55
Feoktistov, Cosmonaut Konstantin,
    19
Ferret satellite, 52–3
Finland, 92
FLTSATCOM (fleet satellite
    communications system), 47

FOBS (fractional orbit bombardment
    system), 19, 20
France, 11, 18, 20, 33, 39, 83, 91, 92

Gagarin, Cosmonaut Yuri, 8, 11
Galosh missiles, 60, 81
Gemini spacecraft, 14, 17, 18, 19, 25,
    33
Germany, Federal Republic of, 20
Glenn, Astronaut John, 11, 12
GPS (global positioning system), 47,
    63, 85
Greece, 26
Grissom, Astronaut Virgil I. (Gus),
    11, 18, 19

Haise, Astronaut Fred, 21
Houston, 21

IDCSP (Initial Defense
    Communications Satellite
    Program), 18
India, 33
Information, need for, 3
Interkosmos Council, 91, 92
International Geophysical Year, 6
International Telecommunications
    Union Convention, 88
Intrepid, 20
ISMA (International Satellite
    Monitoring Agency), 90–1, 94
Israel: Arabs, wars with, 54
Italy, 20

Japan, 18, 19, 20, 33, 39, 83, 92
Johnston Island, 64
Jupiter probes, 23

Kagoshima, 18
Kapustin Yar, 14
Kennedy, President John F., 30, 64
KH-11 satellite, 37–9
Khrushchev, Prime Minister Nikita,
    59–60, 63, 64

100

SAMRO (satellite militaire de reconnaissance optique), 39, 83
Sandal launcher, 14, 32
Sapwood, 14, 32
SAS 1, 20
Satellites:
General references
accidental damage to, 89
aircraft tracking, 44
cameras on, 15, 36, 50
computers, 51
crisis monitoring, 23, 25, 40, 41, 42, 53–5, 91
digital data, 51
digital transmission, 37–9
disarmament proposals and, 92–5
disarmament treaties, verification by, 4–5, 53, 88, 90–1, 93
domestic spying and, 39
false colour photography, 51
film capsules, recovery of, 8, 10, 37, 38, 50, 51
firsts, 5, 8, 10, 11
geostationary orbit, 44, 46
groundtracks, 24, 26, 41, 54, 56
ground stations, 67, 70, 71
'housekeeping', 67
international cooperation, 91–2
military and, 1, 3–4, 30–2, 36, 49–50, 82–8
missile accuracy and, 47, 84, 86
missile test monitoring, 40, 44, 53, 84
nuclear reactors, 43–4, 52, 84
numbers launched, 11–12, 15, 36, 94
orbital life, 23, 34, 37, 55
orbits, characteristics, 16–17
photography techniques, 50–2
propaganda and, 6
radar on, 43, 52, 84
search and rescue project, 92
sensors on, 3, 15, 37, 44, 49, 55, 84, 85 *see also* cameras on
ships, tracking, 40–3
'sleeping', 67
'spare', 94
submarines, detection of, 40, 84
submarines, navigation and, 3, 47
technology, 50–2

types and roles, 34–48, 83–6
weight, 9, 39
Individual countries
Canada, 11, 92
China, 19, 39, 83, 92
Finland, 92
France, 18, 20, 39, 83, 91, 92
Italy, 20
Japan, 18–19, 20, 39, 83, 92
NATO, 20
Norway, 92
UK, 20, 92
USA:
communications, 10, 18, 25, 46–7, 84–5
early warning, 10, 44, 63, 84, 89
electronic reconnaissance, 39–40, 53, 55, 83–4
geodetic, 85
navigation, 10, 47, 63, 85, 88
nuclear explosion detectors, 15, 17, 44, 88
ocean surveillance, 40–3, 49, 84
photoreconnaissance, 8, 12, 15, 25, 35, 36–9, 50, 55, 83
weather, 10, 63, 85–6
USSR:
communications, 18, 23, 46–7, 66, 84–5
early warning, 44–6, 84
electronic reconnaissance, 25, 39–40, 52–3, 55, 83–4
geodetic, 85–6
hunter-killer, 1, 15, 56–7, 60–4, 87
navigation, 47, 66, 85
ocean surveillance, 40–4, 49, 52, 84
photoreconnaissance, 23, 25, 26, 35, 36–9, 53, 54–5, 66, 83
weather, 85–6
see also ASAT *and under names of*
Savitskaya, Cosmonaut Svetlana, 27
Scarp missile, 32, 43
Schirra, Astronaut Walter, 12, 20
Scott, Astronaut David, 18
SDS (Satellite Data System), 46
Sea Lite, 77
Seasat, 52
SEESAW, 69